For the Love of Rivers

For the Love of Rivers

A Scientist's Journey

Kurt D. Fausch

Illustrations by Kristine A. Mackessy and Shigeru Nakano

Oregon State University Press
Corvallis

Library of Congress Cataloging-in-Publication Data

Fausch, Kurt D.
For the love of rivers : a scientist's journey / Kurt D. Fausch ; illustrations by Kristine A. Mackessy and Shigeru Nakano.
pages cm
ISBN 978-0-87071-770-3 (paperback) — ISBN 978-0-87071-771-0 (e-book)
1. Stream ecology. 2. Stream health. 3. Human ecology. 4. Nakano, Shigeru, 1962-2000. I. Title.
QH541.5.S7F38 2015
577.6'4—dc23
2014041829

∞ This paper meets the requirements of ANSI/NISO Z39.48-1992 (Permanence of Paper).

First published in 2015 by Oregon State University Press
Third printing 2024
Printed in the United States of America

Oregon State University Press
121 The Valley Library
Corvallis OR 97331-4501
541-737-3166 • fax 541-737-3170
www.osupress.oregonstate.edu

To Debbie, Emily, and Ben—my family

Contents

Introduction

All of us are profoundly shaped by the boundaries in our lives and in our world. Our language, our culture, our location and generation all keep us grounded in one place and one time, linked to family and friends and home. As a scientist, I have often visited and worked for some weeks in foreign countries, where my boundaries immediately become clear. I cannot travel, cannot find food, and cannot understand local events until I begin to learn enough of the new language to have brief conversations, and can find someone who speaks a bit of English. I need an interpreter to cross the boundaries of language and culture, because transcending these yields vast possibilities for new experiences and learning.

And, like this experience of crossing boundaries of language and culture in foreign countries, it is often only in attempting to transcend other boundaries in our work and lives that we gain a deeper understanding of the world that sustains and nurtures us, and of our own selves. This book is the story about my journey to cross boundaries in nature, the science of ecology that seeks to understand nature, and in my personal life, to discover what is essential for humans about rivers and streams. It is a story about a career-long quest to discover the nature of streams.

Streams are deceptively simple ecosystems. Nearly every stream I ever sought secrets from has been relatively narrow and shallow, and could be easily waded across. Most had clear water, and from the right vantage I could see to the bottom in all but the deepest pools. The first one I attempted to study in detail is a small headwater branch of the Au Sable River in Michigan, a tributary to a river long famous for its trout fishing. Over the course of about four weeks of intensive field research, my first in graduate school, I studied the behavior of native and nonnative trout to determine what habitats they compete for, and who wins. But the more important truth I discovered was that streams are a deep and complex habitat, connected with other ecosystems in all directions. They rival coral reefs and tropical rainforests in color and richness of life. Streams are deeper than you think.

This work was the beginning of my long association with streams, their fish, and the animals that live in the riparian forests and grasslands along their banks. And it was the start of a long collaboration with many fish biologists and stream ecologists who seek to cross physical and conceptual boundaries to understand these interconnected habitats and the web of species they support. It led me from Michigan to regions throughout the western United States and parts of Canada, and to extensive work in Hokkaido, the northern island of Japan.

But gradually during this long association, I also began to learn from the writings of other scientists and to see firsthand that streams and the animals they support are in trouble. For example, between one-third and three-quarters of all the species of freshwater fishes, amphibians, crayfishes, mussels, and snails in the United States are imperiled. Their names are entered on the lists kept by various state and federal government agencies and nongovernmental organizations of species that are now so rare that they are at risk of extinction. Many of these animals are being lost because we damaged their habitat, removed most of the water from their rivers, or simply prevented them from moving among the different habitats they need to complete their life cycles. As Aldo Leopold, our greatest conservationist in North America, explained, every ecologist, including those of us who study streams, "lives alone in a world of wounds." The loss we see is often beneath the surface, invisible to others. Too few others can cross the boundary.

So this is a story about a personal journey of discovery, both in science and in my life, of which this science is only a part. And even though the story is about science, this is not a science book. Rather than simply explaining the science of streams and fish and the ecological relationships that bind them together, my goal is to draw you into the world of streams and their riparian forests and the animals supported by these linked habitats. More importantly, I want to show you the personal experiences of the scientists who study these animals, and these places. I want to draw you across the reflective boundary, *show* you the science, how it is done and why people do it, rather than simply reporting the results to you.

Through this journey of more than three decades, I discovered new places, new people, and new streams, especially while working with Japanese fish biologists in the remote mountains of Hokkaido Island, the northern island of Japan. Together we uncovered new truths about the beautiful Japanese charr that inhabit these remote streams, and about the streams themselves. We forged new friendships and taught and encouraged other new scientists in both our countries who would carry on this work. We lost our closest Japanese colleague, Shigeru Nakano, in a tragic accident at the peak of his life and career, and supported each other and his family in recovering as best we could from this loss. We followed the trail of clues he left through his research legacy, and these led to new discoveries that helped us triumph over this tragedy.

But throughout this journey I also discovered firsthand the tragic loss of streams and their animals, the invisible world of wounds in both Japan and my home region in western North America. And this caused me to ask what it is about streams that is essential for us as humans, and what fundamental steps will be needed to sustain real streams on our planet. I wondered, what would cause people like you and me to care enough to stem this destruction and loss, and attempt to restore streams?

Ultimately, I realized that, like trees and music and good health, streams and rivers are a gift to us as humans. They are part of the essence of our lives in these surroundings we inhabit, within our own boundaries of place and time. They cause our deepest emotions to gather and flow near the surface, and make our days and our seasons and years worth living. They are essential for us to be whole people, psychologically and spiritually. We need their sounds and their views, and their sound advice. And, in the end, I believe we will need to understand how and why we love rivers, if we hope to conserve them.

No book can be written without the support of many people and organizations. Funding for much of this project was provided by the Ecology Program and the Japan Program of the National Science Foundation (NSF), with additional support from Colorado State University for a sabbatical leave. I thank Anne Emig, Alan Tessier, Saran Twombly, and Penny Firth of NSF, as well as Emily Bernhardt and several other anonymous reviewers for encouraging me and seeing the value in this work. Writer's residencies at the Sitka Center for Art and Ecology in coastal Oregon and the H. J. Andrews Experimental Forest in the Cascade Mountains provided nurturing environments for reading and writing, and these were facilitated by Fred Swanson, Sarah Greene, Eric Vines, and Jalene Case. The Leibniz Institute of Freshwater Ecology and Inland Fisheries on the outskirts of Berlin, Germany, also provided a stimulating location for writing, during a visit invited and hosted by Klement Tockner and Robert Arlinghaus.

I thank many friends and colleagues for thoughtful advice and gratifying praise and encouragement during difficult times, all of which have been needed to develop the confidence to complete this book. These include Jeremy and Dana Monroe, Kathy and Frank Moore, Colden Baxter and Emma Rosi-Marshall, Fred Swanson and Gordie Reeves, Hiram and Judy Li, Stan Gregory and Kathryn Boyer, David Noakes, Mike Young and Bruce Rieman, Jason and Susie Dunham, Joe Ebersole, Ellen Wohl, Bill Timpson, John Daniel, Pat Wheeler and Diane Cook, Sue Ellen Campbell and John Calderazzo, Brian Richter, Chuck Crumly, and Mark Caffee. I also thank my friends, family, and colleagues who read chapters and offered comments and constructive critiques, including Colden Baxter, my late colleague Bob Behnke, Kevin Bestgen, Dan Bottom, Charley Dewberry, Jeff Falke, Deborah Fausch, Mikio Inoue, Chris Jordan, Daisuke Kishi, Satoshi Kitano, Boris Kondratieff, Fabio Lepori, Yo Miyake, Jeremy Monroe, David O'Hara, Doug Peterson, Bruce Rieman, and Julie Scheurer. I offer my special thanks to Yoshinori Taniguchi, who read several chapters and answered countless questions about specific details of past events and Japanese language, culture, and custom. To all, this book is far better for all your advice, work, and support.

Jeremy Monroe and Dave Herasimtschuk of Freshwaters Illustrated were companions and advisors throughout this journey, including trips to Japan and Colorado, creating videos and images that set the book apart, and advising on illustrations and design and promotion. Kristine Mackessy proved a talented and patient illustrator, turning my visions for illustrations into the reality of pen-and-ink drawings, and Erin Greb provided skilled expertise in creating the maps. To all these artists and friends, this book owes much to all your creativity and efforts and help, which far exceeded all I could have hoped for.

I extend a most special thanks to Mary Elizabeth Braun of Oregon State University Press, whose encouragement and insight and vision lifted the book above what I could have imagined. Susan Campbell provided not only able copyediting but also welcome encouragement as a kindred spirit, and Micki Reaman, Marty Brown, and Tom Booth of the press provided assistance in production and marketing.

And, finally, I owe a great debt to my wife and friend, Debbie Eisenhauer, through our more than four-decade love affair. You have been a patient and understanding supporter, who believed in my work, ran many errands, made countless meals, mentored children during my frequent short and long expeditions to the field, and managed family crises so that I could do all the research described in this book, and write the book itself. My greatest triumph is to have shared with you the joys of filling each other's cup, but not growing in each other's shadow, and together gladly bending in the archer's hand so that our children could go swift and far.

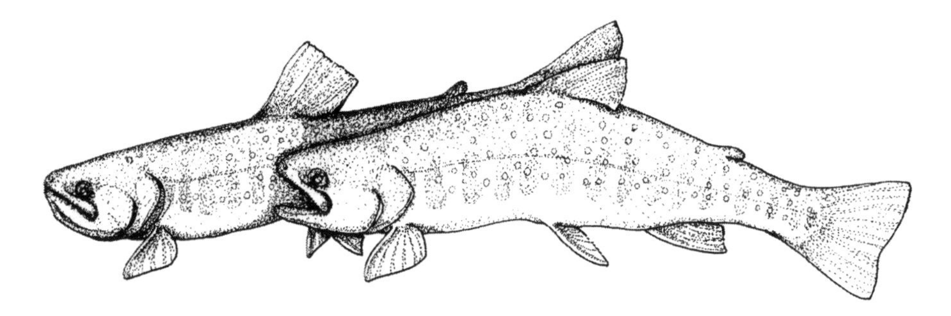

Illustration by Shigeru Nakano

Chapter 1
An Awakening

I found that my life had changed the first time I crossed the reflective boundary to look beneath the surface of a stream. It must have been early summer 1977, judging from the old photographic slides, on the Salmon Trout River in northern Michigan. During a reconnaissance trip, my graduate advisor Ray White suggested that we explore the stream by snorkeling, as a way of helping me gain firsthand experience in the environment of the trout I would soon study. I spent several hours underwater, enthralled by what we saw, until I was too cold to dive anymore. On that day, the door to a new world had opened.

Although the stream there averaged only about knee-deep, and was nowhere deeper than my waist, the view was of a place much deeper and more complex than I had imagined from above. Everything is magnified about a third larger underwater. The undercut stream banks looked darkly cavernous and eerie, and even small fish appeared gigantic at first. Rocks that were dry and dull lying beside the stream became deep reds and blues beneath it. The currents and bubbles formed artful patterns on the silver ceiling of the water surface above. An underwater wind created by the relentless flow waved the long feathers of aquatic plants and buffeted the short wisps of neon green algae that clung tenaciously to bare rock.

But these rocks and plants were mainly a backdrop to the beauty and mystery of the main players that amazed me in this riverine drama, the aquatic insects and fish. Every new vantage revealed more members of an intricate underwater community. Mayfly nymphs scurried across the surfaces of rocks, stopping intermittently to mow down the thin lawn of algae. Caddisflies built and tended miniature fishing nets to collect bits of chewed-up dead leaves discarded by other invertebrates upstream. Deep green brook trout burst from their resting places to capture aquatic insects drifting along, or hapless terrestrial insects that fell onto the surface of the stream. These fish looked so different from the same creatures that flopped awkwardly in my hands on the stream bank after we captured them during our sampling. Underwater, when I met them at their level, they moved so gracefully, and were so well suited to their flowing liquid environment.

I first realized then, without forming any clear thought, but only feelings, that this was a world I was drawn to understand. This was a place worthy of focused study, and these were creatures that could fascinate me for a lifetime. Nearly every person I have met since then who has crossed this boundary to explore beneath the surface of a river or stream has emerged with a similar surprise and fascination. We humans are aliens here, unable to breathe the atmosphere or move through it easily, or even see clearly when we open our eyes underwater. But by the simple technology of mask and snorkel we are allowed to at least visit as clumsy foreigners and, in those brief sojourns, to encounter vistas of transforming beauty. Unfortunately, to most people this world is unknown, hidden beneath a mirror that allows only rare glimpses when seen from just the right angle. I felt privileged by the chance to venture further. Where would the journey lead?

‹ ›

In early summer 1991, I came to the mountains of northern Japan to cross this boundary again, and seek with a new colleague some of the scientific truths hidden beyond. In the fourteen years between, I had completed several studies of fish by snorkeling and had mastered many techniques and bought better gear. But in this season the snows were still melting at the headwaters, and the cold from the 45°F water crept through the two layers of fleece I wore under my dry suit after only twenty minutes of lying completely still in Poroshiri Stream. The tight rubber seals at my neck and wrists, the confining neoprene hood, and the cold-hardened mask pressed against my face all made me more uncomfortable. Straining my neck to look straight ahead while lying on my stomach made it even worse. And yet, despite all the discomforts, the fish were exquisitely beautiful, as the grand ecological play unfolded around me.

Dolly Varden and whitespotted charr are the two trout native to this remote mountain stream in Hokkaido, the northern island of the Japanese archipelago. Dolly Varden are called *oshorokoma* in Japanese, a name of unknown meaning given by the native Ainu people who originally inhabited this island. Whitespotted charr are called *iwana*, which means "rock fish." Both charr are a deep forest green with light-colored spots on their sides, and look superficially similar. Dolly Varden's namesake is a fashionable young woman in Charles Dickens's 1841 novel *Barnaby Rudge*, who inspired brightly patterned dress fashions in the 1870s. This fish is easily distinguished by the small brilliant crimson spots among the dark thumbprint-shaped "parr" marks along its sides. By contrast, the sides of whitespotted charr are covered by many large white spots that to me looked unreal, larger than those on any trout or salmon I had seen in North America.

The eight charr I was studying in this pool cruised leisurely in the wakes just behind and slightly above rocks sticking up from the streambed, each finding a refuge there from the full force of the flow. The currents buffeted them slightly up and down and side to side as they swam, like kites riding in the wind. But no fish strayed far from its position in the carefully choreographed pecking order spaced throughout the pool, a dominance hierarchy they tested every morning. Each scanned the on-

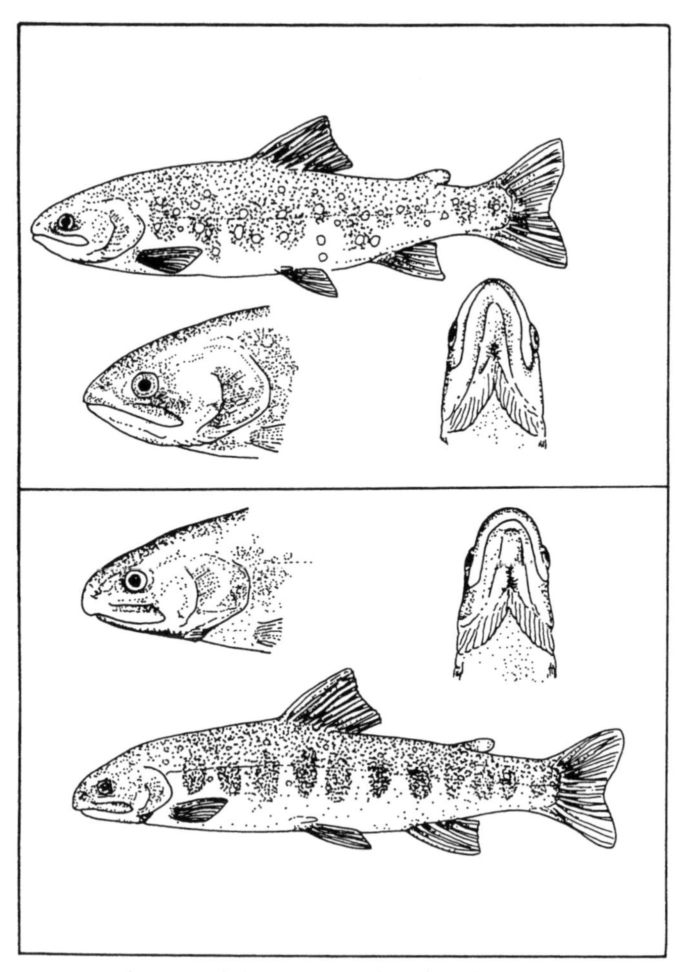

Figure 1. Whitespotted charr (top panel) and Dolly Varden charr (bottom), showing spotting patterns and parr marks, mouth position, and the lower surface of the head. Illustration by Shigeru Nakano (used with permission).

coming current for drifting food. Then about every ten seconds one fish or another gave a startled flick of its tail and stopped swimming for a second, a sign that it had noticed a tiny drifting aquatic insect. Turning slightly to focus intently on its prey, it sprinted to the side, engulfed it, and returned quickly to its rightful position so as not to upset the grand order. Other charr rose to the surface to snatch larger terrestrial insects that had stumbled and fallen from the forest above. Once every five minutes or so, neighbors in the dominance hierarchy squared off and fought over the borders of their territories with ritualized displays, like sparring

Japanese martial artists. Score settled, and the order usually unchanged, they returned to their feeding.

In my first research underwater, fourteen years earlier in Michigan, I had wanted to study firsthand whether nonnative trout drove native trout from the best feeding and resting positions in streams. Like most graduate students starting out, my topic and approach, my hopes and dreams, were strongly influenced by my fellow students. One of these, Charley Dewberry, introduced me to the interesting problem of nonnative trout invasions and helped me plan how I could study it directly by measuring these fish positions while snorkeling, an approach only a few ecologists had used in streams. The insights from that first study set the course for much of my later work. In 1991, early in my university career, I joined forces with a young Japanese colleague, Shigeru Nakano, to pursue a new dream in this unlikely place. Our quest was to explore the mystery of these charr in remote headwater streams in the mountains of central Hokkaido, a land of salmon and brown bears in northern Japan.

☾ ☽

Nakano moved confidently, effortlessly, between land and water during our field research, that first summer and the next. On the first day of our work, he showed how to wrap fist-sized stones with brightly colored plastic flagging and label them with letters and numbers, placing them in a grid on the streambed as references for a detailed map of the stream pool. Working quickly, he waded into the stream and drew outlines of the main rocks and boulders in each grid cell on plastic graph paper. Slipping easily into his dry suit, mask, and snorkel, he then entered the pool to compare the map with the underwater view, and adjusted it to highlight the most important stones and features that a snorkeler would notice when observing fish from a downstream vantage.

Fingers worked quickly to attach the fine monofilament line to the end of his telescoping graphite *tenkara* fishing rod. *Tenkara* fishing, the traditional style developed for small steep mountain streams in Japan, originally used bamboo cane poles to place flies or bait in precise locations in the plunge pools. Nakano had grown up fishing this way, and had adapted the technique for research to allow fishing while snorkeling. With deft motions, he attached a very tiny hook with no eye to the end of his line, using many wraps around the shank to create an intricate snelled knot. Holding a small hand net downstream, he kicked up some streambed gravel to dislodge aquatic insects, and peering into his catch he picked out a large caddisfly larva to thread on the tiny hook. Clad in his dry suit and rod in hand, he crawled into place at the downstream end of the pool and then, looking underwater carefully, laid the bait into the current a few feet upstream of the most dominant fish's position. The large prey drifted naturally into view, and the fish rose to capture it. Nakano tightened the line when he saw the strike underwater, and the fish was easily caught with little effort or harm. He smiled as he used the hand net to capture the fish and prevent it from struggling further, and rolled out of the pool to remove the hook and mark the charr for future identification. A pair of tiny plastic colored ribbons sewn just beneath the skin on the fish's back trailed in the current when it was released and identified the fish as Red-Green when observations of its behavior were recorded on the map throughout the summer.

Only a few times in the career of any ecologist, all the stars and planets line up and a carefully planned field experiment actually works. Most often things go wrong, and at best the experiment answers a different question than you asked. Failures are also common in laboratory experiments, but there is usually the chance to repeat them. Go get more fish and start again. In contrast, field experiments in ecology, especially stream ecology, are so much work and so expensive that they rarely can be repeated. First, they usually can be done only once a year, often in summer when fish and invertebrates are most active and grow rapidly. Then, mesh fences or cages must be erected to enclose the fish or invertebrates. But as soon as these are in place, pieces of leaves and sticks carried by the stream quickly begin to clog the mesh, which then must be cleaned frequently throughout the whole experiment. Otherwise, the stream will rip, flatten, or wash away the structures, or carve great gaps in the gravel bed or banks to which they are staked. So field experiments often become an ordeal, creating a kind of history for us, and freezing memories about our colleagues and ourselves in a particular place for particular moments in time.

On the last day of our two-summer expedition, in July 1992, we needed this last trial of the field experiment in Poroshiri Stream to work. The six large fine-meshed nets we had placed to filter drifting insects from the

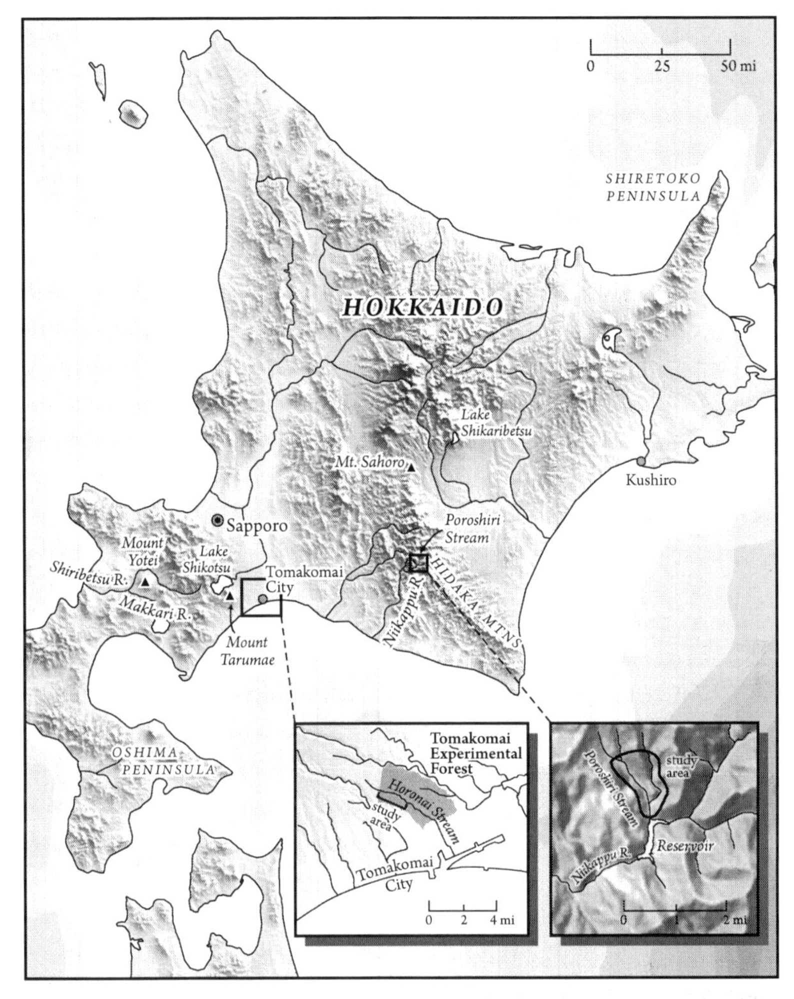

Figure 2. Map of Hokkaido Island, the northern island in the Japanese archipelago (see inset of figure 3). Insets show locations of field studies in Poroshiri Stream in the Hidaka Mountains (right inset) and Horonai Stream in the Tomakomai Experimental Forest (left). Other locations of research are in and near the Makkari River basin and in Shiretoko Peninsula.

entire stream flowing into this pool were quickly clogging with leaves and sticks. The water was backing up behind them and would gradually overtop the nets, even with frequent cleaning. Would the nets tear under the pressure, or would the weight of the water bend the metal stakes that held them and break the small streamside trees to which we had tied the stakes for support? To add to my worry, rain began falling, a steady cold rain stronger than a drizzle.

"*Ame ga fute iru*," commented Satoshi Kitano, my Japanese graduate student assistant, followed by his perfect, even-toned English translation, "It is a rainy day today." Diving in my older wet suit for the past four days had been brutally cold, after the rubber neck seal of my dry suit split from top to bottom. We had worked many weeks in a row, and dove most days, to record underwater observations of fish and their behavior. When I entered the stream in my wet suit, the first fingers of cold water grabbed my chest and back, taking my breath away and making me gasp through my snorkel. I could tolerate only short twenty-minute dives separated by breaks for hot tea, chocolate candy, and jumping jacks to stop the shivering. But Nakano-san had worked wonders by fixing the neck seal of my dry suit with a large tire patch when he returned to the field from his short stint of summer teaching at Hokkaido University. Now I could make hour-long dives again in relative comfort. These logistical problems and discomfort were not new. I had learned to expect them from field research on stream fish.

Carefully entering the cold water again at the downstream end of the pool, I inched forward on fingertips and settled into the uncomfortable position I would hold for the hour. I punched my watch button to start the five-minute waiting period, allowing the fish to settle into their locations and resume normal behavior. After marking and labeling positions of all the charr on clear plastic graph paper laid over the map that Nakano had made, I set to recording the feeding forays and ritualized contests of individual fish for five minutes each.

But just there! What was that? A small 5-inch Dolly Varden low in the pecking order near the downstream end of the pool began to switch its feeding behavior. Then others followed suit. Just as we had predicted from Nakano's previous research, instead of scanning the flow and darting into it to catch drifting insects, they pointed their whole body downward and lunged at rocks on the streambed, dislodging tiny mayflies and stoneflies that scuttled over the rock surfaces. One by one, from the bottom of the pecking order up, as the supply of drifting insects was filtered out by the nets upstream, they gave up feeding on this declining "drift" and switched instead to "picking" insects from the bottom. Some began traveling in wide figure-eight patterns along the stream bottom, circling up to 4 feet to each side of their former position, searching for prey. Some even wriggled beneath the cobbles and small boulders like snakes to look for insects and popped up farther along their paths, ignoring the artificial boundary that we humans call the streambed.

"It worked!" I exclaimed, speaking the words softly but excitedly in my throat while clenching the mouthpiece of my snorkel. I expected no one to hear but myself, and hoped not to scare the fish. It *had* worked. Our scientific prediction had been confirmed, and more importantly, the four trials of the field experiment I designed to test it had not been flooded and our nets washed away, or otherwise compromised and ruined. We had crossed the boundary from our home in the terrestrial world to wrest some small but important truths about these beautiful fish from beneath the shiny stream surface that masks our vision. The years of preparation and grueling work had paid off. Even hours later, when all the tedious observations were finally complete and the data recorded, I literally jumped for joy and yelped like a rodeo cowboy when I emerged from the water.

Although I couldn't imagine it then, this unlikely success signaled only the beginning, the first of many milestones in a journey undertaken by Shigeru Nakano and me that crossed into far country and followed unexpected openings in our work and lives. In the end, we changed the way that each other did science. And the research that grew from these beginnings ultimately led to understanding new ways that streams are inextricably linked with the forests and grasslands they traverse. Our friendship flourished, rooted firmly in our common passion for field research and a deeper understanding of streams. The journey not only transcended boundaries of language and culture and the sheer distance between our two homes, but also encountered tragic loss, and gradual renewal. Despite this loss, these experiences enriched my life, gave my work more meaning, and helped me develop a more mature view of science and the world it seeks to reveal. And they also forced me to consider the future of streams, and why humans need them.

‹ ›

Roofs of Japanese homes built in the traditional style, with glazed tiles of dark brown or blue, emerged from the patchy mist beyond my window in the Boeing 747 as we descended toward Tokyo's Narita Airport. Each home was surrounded by rice fields, now just harvested at the end of the season. My awakening to Japan began on this day in October 1988, four years before our field experiment in Poroshiri Stream. And despite the eighteen hours of traveling on buses and planes in cramped spaces (I'm 196 centimeters in Japan, 6 feet 5 inches in the U.S.) and the dull fatigue, I had never felt so exhilarated by an opportunity. The bump at landing and the long, harshly lit corridors led to the bustling baggage claim and passport check before the scavenger hunt to find the connecting flight to Sapporo among the warrens of an entirely foreign airport. I struggled to avoid falling asleep with my eyes open while I waited for the next flight in the hot and crowded waiting room. I wasn't yet used to this kind of world traveling.

Why did I go to Japan in the first place? To be honest, I didn't expect to learn very much new from Japanese scientists about my field, the ecology and management of stream fish. American scientists in this discipline tend to teach in departments titled "Biology," or "Fisheries and Wildlife," or conduct research for federal agencies like the U.S. Fish and Wildlife Service. They see their calling as understanding the lives of stream fish well enough to be able to conserve the habitat in rivers and streams that sustains them. Many study the popular trout, salmon, or bass, but others work on little-known fish like minnows, suckers, and even sturgeon and eels. Eventually, I worked with graduate students and other scientists to study both the popular and the unknown.

I had received a letter inviting me to present my research at an international conference on "The Biology of Charrs and Masu Salmon" in Sapporo, the capital of Hokkaido. I was also asked to serve as chairperson for a session of other speakers, paired with a Japanese scientist whom I didn't know. Only six years into my academic career at Colorado State University, I was a relatively unknown scientist. My work had never received enough notice to prompt such an invitation, especially not from overseas. I was excited for the chance to meet famous fish biologists from Canada and Scandinavia who studied these charr and whose writings had inspired me.

My colleague Bob Behnke, a fellow professor and world-famous expert on trout and salmon, was also invited. He had told me stories of traveling into the mountains of northern Japan to catch beautiful charr as a soldier on R&R during the Korean War. I dreamed of seeing these exotic new fishes for myself. Being in Bob's company made me feel more comfortable about attending the foreign conference, and I gladly accepted the invitation by return mail, in those days before the Internet. But I knew of only one Japanese scientist in the field, Dr. Hiroya Kawanabe, who was the leading host of the symposium, and this was based on only one paper he had published nearly twenty years earlier. I didn't think the Japanese had done much other research that would interest me.

Takai, ne! (Tall, isn't he!), shouted the junior-high school boys following me along the busy Sapporo city street at night, awash in the light from garish neon signs. They ran and leapt into the air behind me, trying to reach my height, apparently to see what being that tall might feel like. Suddenly I realized that there are few people as out of place as me in Japan. The beds are short and the doorways low, but at least this makes the polite custom of bowing when one greets and leaves people a natural tendency for me. In fact, it's a necessity for self-preservation. Fear of running into a door-closer right between the eyes is a great incentive. But these inconveniences were far outweighed by my natural curiosity and interest in the food and the environment of Japan, and the warm hospitality of Japanese people.

Hokkaido is blessed with abundant seafood, combined with fresh vegetables and meats that are more similar to the midwestern United States than any other part of Japan. Fertile fields grow potatoes and carrots, and grass for dairy and beef cattle. Autumn is the season when they serve a special dish, *ishikari nabe*, a soup of salmon and vegetables boiled in a pot right at your table. Because the islands of Japan are small, most food is local by necessity and most of it is extremely fresh. Although many Americans are squeamish about eating raw fish and squid and various kinds of seaweed, I was fascinated by the amazing diversity of fresh and natural foods in Japan. When we encountered a strange new food in a restaurant, Nakano often asked me with some surprise, "Is it all right for you?" I rarely declined.

I had worked many weekends and nights through the late summer and early fall of 1988 in Colorado to collect and prepare the field data that I would present at the Charr Symposium. Days of sampling mountain streams in rough country were shoehorned between days of teaching undergraduate courses and mentoring graduate students. I had no funding. So I worked with student volunteers, or alone when I found none (not advisable), packing field gear and people into our tiny family car and driving it over rough mountain roads. The work was grueling, and I fell sick after one particularly long day. A shortcut back to the trailhead turned into a harrowing hike over a ridge through a trackless forest. I had pushed myself and my volunteers too hard. Every waking hour on nights and weekends that my family could spare me I analyzed my data, and compared it to data of others scientists that I reanalyzed. So after arranging guest lectures in my courses to allow the two-week trip to Japan, I was excited to finally reach the meeting and present my new ideas.

Nonnative trout and charr introduced around the world over the last 135 years, such as brown trout from Europe and rainbow trout from our Pacific coast, have invaded many waters and pushed out native trout. For example, brook trout (which are actually a charr), native only to north-central and northeastern North America, were introduced to streams and lakes throughout the Rocky Mountains and invaded upstream into many headwaters until they reached impassable waterfalls or cascades. They wiped out many populations of the colorful native cutthroat trout that define the West (as I explain in chapter 7), and often left only small remnant populations above these movement barriers. However, I found several places in Colorado, and reports on other places in Montana and Idaho, where brook trout invasions had simply stopped as streams grew steeper, apparently creating refuges farther upstream for the native cutthroat trout, even without barriers. I wondered what factors determined why brook trout invaded some stream reaches and not others, and what prevented them from invading farther upstream? These findings are important for preventing extinction of the native trout, and I was excited to hear the reaction from other more experienced fish biologists at the symposium. I had no idea that what I would learn there would change my entire career.

I peered close to marvel at the brilliant crimson and red-orange colors in the fins of the charr and on their bellies and the spots along their sides. The underwater images taken by Japanese photographers and hanging in the meeting room on the campus of Hokkaido University were not just beautiful, they were stunning. During this first evening mixer I mingled, but I knew personally almost no one at the conference. So, being rather shy, I stood aside and attempted to blend into the wall, an oxymoron for someone of my height. Suddenly, someone I took for a Japanese graduate student, a bit bulky and wearing a striped rugby shirt, strode toward me. He had an older professor in tow, dressed in a suit. "Hi! Are you . . . Fausch-san?" he asked hesitantly, with a rather husky voice for a Japanese. "Yes," I replied as hesitantly, surprised that this young man would know my name. He continued in awkward English, "My name is Nakano Shigeru—here is Dr. Ishigaki Kenkichi," using the tradition in Japan of listing their surnames first before their given names. "I read your papers. We wanted to show you Dr. Ishigaki's recent book, *Exploring the Mystery of Charrs.*" I was immediately intrigued. What was in this little yellow book? How did this young man from so far away know about me and my work? By that time, I had published only a few papers myself, and they had appeared only a few years before.

Since the earliest glimmers of ideas in biology, a field that started only around 1800, explorers whom we would now call biogeographers have wondered what explains why plants and animals occur where they do on the planet. For example, Alexander von Humboldt, an early German explorer who made a five-year-long voyage to Central and South America to collect specimens, described gradual changes in the flora as he climbed to the peak of Tenerife, the largest of Spain's Canary Islands, in 1799. Although the island is in the subtropics off North Africa, he reported that plants near the top were similar to those found much farther north in Europe. Marine ecologists like Joe Connell pondered why one barnacle species replaced another at higher tide levels on seashores worldwide. Nelson Hairston Sr. spent years working out why salamanders replaced each other at different elevations along north- versus south-facing slopes of the southern Appalachian Mountains. Do different species simply tolerate different temperatures, or periods of drying at low tide, or do some species dominate the best spaces and force others to live at the margins? Can plants and barnacles and salamanders actually compete with each other? Even more intriguing, could changes in temperature or humidity caused by elevation also change the competitive ability of these species, allowing one to win in certain elevation zones but forcing it to lose in others?

I had been fascinated by these questions from my earliest days in graduate school at Michigan State University. I started by studying why nonnative brown trout had replaced native brook trout in so much of their former habitat in Michigan and other Great Lakes states. In most watersheds, brook trout persisted only in small cold headwater streams, whereas brown trout had invaded the warmer mainstem rivers. Had logging allowed the rivers to warm too much and simply made the habitat unsuitable for brook trout? Or did brown trout actively exclude them by taking over prime feeding and resting positions? I spent days living like a mole in the library basement, scouring the literature about trout and salmon competition.

That summer of 1977 I conducted my first large-scale field experiment, in which I removed brown trout from more than a mile of stream and snorkeled before and afterward to measure whether brook trout shifted their stream positions in response. The results clearly showed that brown trout could exclude brook trout from prime resting spots. I desperately wanted to do more research on this topic for the native and nonnative trout in Colorado but was stymied by a lack of interest from agencies with funding. But here in Ishigaki's book were the interest and the clues for charr in northern Japan.

Nakano opened the small book, written entirely in Japanese, and my eyes grew wide in amazement. The tiny intricate Japanese *kanji* characters really were arranged in straight lines flowing from the top to the bottom of each page. And the book really was read backward, just like we had all learned in grade school. "Dr. Ishigaki write this book from his PhD research—he studied Dolly Varden and whitespotted charr in Hokkaido," Nakano struggled, using unfamiliar English. He flipped to near the end of the book, past maps showing the many dots where the two charr had been captured in Hokkaido watersheds, and pointed to a graph. "In this figure he shows Dolly Varden are always upstream, and whitespotted charr always downstream. But they both shift higher up in mountain streams as we go southwest across Hokkaido. He also did simple lab experiments to see if whitespotted charr always dominate Dolly Varden."

I was stunned. For my talk at this conference, I had scoured the literature written in English to find the few papers that described these kinds of distributions of trout and charr in North America and Europe. But here was a whole book on the subject. My mind raced. What could explain why the charr shifted to higher elevations across the island? How could one answer that question? Almost no one was interested in my ideas in Colorado, yet here was a senior scientist who had laid all the groundwork in northern Japan. And all this was going on far beyond my worldview.

Dressed in casual clothes and speaking freely to others, Nakano was entirely out of place among the polite and reserved Japanese scientists, many of whom dressed in suits. The weak podium light had bothered most speakers at the symposium, making it hard for them to read their notes, and added to the language barrier that many suffered. "Oh—I will just use my field headlamp," Nakano announced as he donned it with a dramatic flair, finding a comical way to break the Japanese tradition of modest decorum. Then he proceeded in his talk to use the theory I had proposed in one of my first papers, about how trout and salmon compete for profitable feeding positions in streams, to explain the dominance hierarchies he had measured between whitespotted charr and masu salmon. Masu salmon are closely related to our coho salmon in the Pacific Northwest.

I was again floored. Who *was* this guy? I had struggled for three years to publish this theory that Nakano was using. My paper was rejected from a high-profile ecological journal and published in a relatively modest one on the second try, only four years earlier. Honestly, I didn't think many other scientists would ever find it and read it, much less use it directly in their research. The highest honor for a scientist is for someone to test their ideas and eventually replace them with a better theory. To my surprise, this paper now has been used and cited by other scientists more than any I have ever written, and the ideas extended and modified by others to improve them.

"Would you like to take *hiro gohan*, uh . . . lunch?" asked Nakano invitingly. "I would like to show you my data, Fausch-san." He led me quickly through the busy Sapporo cityscape, crowded with small white and black Toyota and Nissan vehicles rushing down the narrow streets. Nakano found a suitable restaurant near a train station with plastic models of the food in a display case out in front, a great help to foreigners like

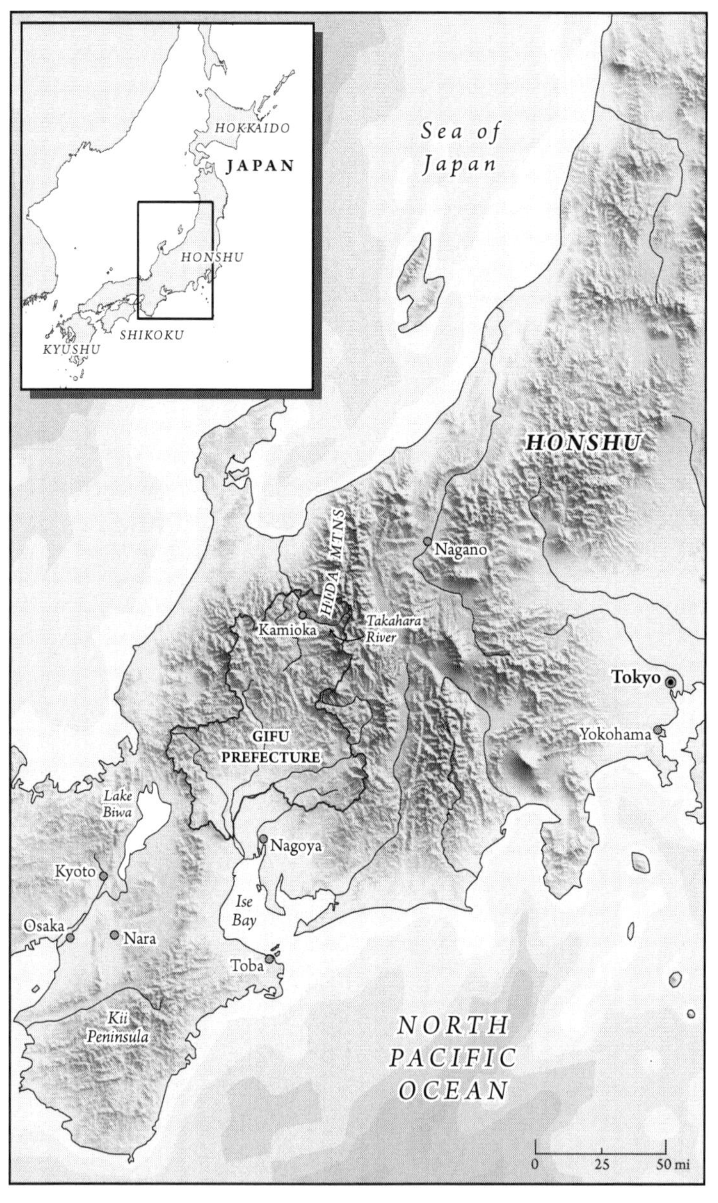

Figure 3. Map of central Honshu Island, the main island of Japan, showing Gifu Prefecture and Kamioka, Shigeru Nakano's home town in the Hida Mountains. The inset shows the four main islands in the Japanese archipelago.

me. Sitting at the low tables and eating with *hashi* (chopsticks) from the many small and elegant dishes of rice, miso soup, vegetables, and broiled fish, he began to show me his amazing work.

For two summers after he earned his master's degree, Nakano had snorkeled for days on end, more than anyone had ever done before, to study the behavior of charr and salmon. He was working as a naturalist in a small museum near his hometown of Kamioka in a beautiful mountain valley of Gifu Prefecture in north-central Japan. When I later saw this region for myself, I knew that the landscape there must have inspired the paintings on ancient Japanese scrolls, of rivers cut deep into forested mountain valleys, bordered by cliffs shrouded in mist. In 1986, Nakano traveled far south to a small mountain stream on the Kii Peninsula southeast of Osaka, where he had previously studied landlocked red-spotted masu salmon (which act like trout) as an undergraduate student. There he made detailed observations of the positions they held and their underwater contests. Snorkeling alone nearly every day for seven weeks, he cataloged winners and losers for nearly 2,200 matches among seventeen different fish observed in three pools. The next summer, in 1987, he dove every day for six weeks in a mountain stream near his hometown, measuring 850 contests among young masu salmon and Japanese charr. All these observations were recorded underwater on his highly detailed maps of the beautiful plunge pools. I had made such maps of fish positions in a small artificial stream in the laboratory during my PhD research, but only a few others had done it in real streams, from towers or the stream bank, and no one had amassed such a large amount of detailed data. His analysis was simple and clear and elegant, and strongly supported several ideas I had proposed in my theory. After he showed me the maps and explained his data, I could only stare at his results, amazed yet again.

Although I had thought I would simply attend the charr conference and see some of Japan on my trip, I had just discovered the best work ever done on trout behavior, by an unknown biologist with a master's degree working in a small regional museum. Not only that, but this person had ideas that were more similar to mine than any scientist I had ever met. "Please send me your manuscripts when you finish them, and I will help edit the English," I offered enthusiastically.

The glow before sunrise filled the V between the forested mountains beyond Lake Shikaribetsu with rose-colored sky, framed above by a thin layer of dark steel cumulus clouds. I shivered a bit as I peered out the small window of the traditional *ryokan* (Japanese hotel) where we had stayed the first night of our field excursion around Hokkaido after the meeting. Everything was so neat and orderly, from each traditional *futon* laid on the beautiful woven *tatami* mat floor to the patterned *yukata* robes used after the hot baths down the hall.

Dr. Hiroya Kawanabe was leading our excursion, one of three throughout Japan, which pleased me greatly, even though Nakano had chosen to attend another one. Kawanabe-*sensei* (meaning "teacher," a term I use here with the deepest respect) had been a striking figure and leader throughout the meeting, dressed in his traditional dark blue Japanese kimono, the academic garb of university professors in earlier times. At each hotel en route, he gave talks about the natural history of fishes in Hokkaido, to large audiences. For example, Lake Shikaribetsu (the "u" is just barely pronounced in names like these, as an afterthought) is home to a unique form of Dolly Varden studied intensively by Japanese scientists. Their gill rakers, like combs in the back of their throat, are longer and finer than other strains of this charr, making them well adapted for filtering out the tiny crustacean zooplankton on which they feed in the lake.

Our tour wound from Sapporo in the southwest, through the beautiful Hidaka Mountains of central Hokkaido, and reached Shiretoko National Park on the farthest northeastern peninsula, which projects into the North Pacific Ocean. Shiretoko Peninsula is Japan's wildest country, inhabited by brown bears and Dolly Varden that migrate to the ocean like salmon. It is a land of steep headlands jutting into the cold and misty ocean, and looks much like Alaska. We toured in a fancy bus, loaded with Americans and hardy Canadians and Scandinavians, along with our Japanese colleagues. Several of the Scandinavians ranged from my height to 6 feet 10 inches, more than 2 meters tall. Japanese cooks and waiters came running outside to see the visiting giants when we stopped for lunch in rural Hokkaido.

But among all that was beautiful in our trip across Hokkaido, among the clear streams that tumbled down rocky mountain valleys and meandered across gentle plains to the North Pacific, I saw much that was not beautiful or natural. Much of the habitat for fish had been destroyed, or blocked. Many gently sinuous streams had been ditched straight through agricultural fields and towns, and many rivers were entirely paved with cement blocks along their lower courses through small cities near the ocean. As a stream ecologist, I know that these changes eliminate the clean gravel salmon and trout need to spawn, and the gravel riffles where aquatic invertebrates live that fish eat. Fish abundance plummets in channelized streams. The banks of other large rivers were lined with huge interlocking four-pronged tetrapods, looking like toy jacks for the children of giants, to control erosion caused partly by the river straightening. Small mountain streams often were blocked by a series of thick concrete *sabo* dams (rhymes with auto) that spanned the entire small valley, apparently to slow destructive landslides in this island of earthquakes and volcanoes. These dams confine charr to short stream segments, preventing their migrations and movements to find the habitats they need for spawning, rearing, and overwinter refuges.

"Why have they destroyed their streams?" I wondered. But shortly after the question formed in my mind I had a realization that would come to me over and over again during my visits to Japan. "Oh yeah,—we also do things like this." After all, we are the country with only one river of any length that is undammed, the Yellowstone River in Montana. Even this river has many water diversions to irrigate hay and other crops, and these barriers also stop fish like trout and sturgeon from reaching the places they need to survive. No culture is alone in damaging streams.

The tour bus stopped along the banks of a river, at a large olive-green slightly rusting fish weir that spanned the hundred-yard-wide channel and shunted upstream-migrating adult salmon into a trap along one bank. The deck of the boat tied up beside the trap was covered with arm-length adult chum salmon, a mainstay for Japan's fisheries and food supply. Chum salmon migrate upstream in the fall to spawn in the lower reaches of Hokkaido rivers. The pink-orange fertilized eggs incubate through the winter in the gravel where the females bury them, and the young fry leave the streams for the ocean in early spring immediately after emerging from the gravel. This means they are ideally suited for being raised in hatcheries. Eggs can be stripped from females, fertilized with "milt" (sperm) from males, and incubated artificially until they hatch and are ready to go to sea. Hatchery rearing eliminates natural predators and diseases that kill most eggs and fry in the wild, so far more chum "smolts" (fry that are ready to survive in the ocean) can be released than would be naturally produced from these rivers.

We had toured the Kushiro hatchery in eastern Hokkaido and seen state-of-the-art technology for eliminating nature's role in salmon spawning. But nature has mechanisms for adjusting animal populations to their food supply, and she was not entirely happy about making room for so many more Japanese chum salmon. Throughout the 1980s, nearly 2 billion juvenile chum salmon were released into the North Pacific Ocean from Hokkaido hatcheries every year. Although the proportion of these that survived life in the ocean and returned as adults rose steadily after 1960 because of better feeding and release methods, the length and weight of these adult salmon declined sharply. Those that returned in the mid-1980s were substantially shorter and lighter than a decade earlier. In short, they ran out of food, even in the apparently limitless expanses of the ocean. Although we now know that ocean productivity also goes up and down over decades-long climate cycles, recent research shows that Japanese hatchery salmon continue to tax the food supply of the North Pacific where they roam and are now likely affecting growth and survival of North American salmon from Alaska.

As our Japanese host described the big green weir from the riverbank, tears came to the eyes of a renowned and otherwise stoic Norwegian biologist who had witnessed the loss of Atlantic salmon in his country from such intensive hatchery operations. "Don't do this!" he cried. "We made this mistake and lost our wild salmon." Japan is now working to seek a better balance.

At the end of the excursion, I winged home to a whirlwind of classes to teach, graduate students to mentor, committees to endure, and two small children to help raise. But memories of all I had seen and experienced in Japan often haunted me. "I should invite Nakano to come study in my laboratory and earn a PhD in the U.S.," I thought. When I inquired, he replied that he had already started on his own in Japan. He had also landed a new position as an assistant professor in an experimental forest of Hokkaido University, located in a very small town in northern Hokkaido. Time passed, yet my desire grew stronger to collaborate with this amazing young ecologist to study charr in Hokkaido.

I realized that, given the rather strict Japanese hierarchical system, if I wanted to create such a collaboration I would need to start from the top. Hiroya Kawanabe, one of the most famous of Japan's ecologists, had been extraordinarily gracious and kind to me during the symposium. So, in late July 1989, I boldly wrote a letter asking if he knew of any sources of funding for such work. My attempt two years earlier to land a grant from the U.S. National Science Foundation (NSF) had failed in the stiff competition, and I remarked that the chances there seemed slim. But Kawanabe disagreed, and replied in early September with a plan for a group of Japanese stream fish ecologists, including Nakano and himself, to join me in applying for funding under a U.S.-Japan joint program he knew about. "This is perfect," I thought. I was sure that Japan's research agency would vote to fund Kawanabe's proposal, given his deep knowledge and many accomplishments. I felt I might then have a chance with the NSF, if I worked very hard to at least get my proposal "in the ballpark." In early summer 1990, I was overjoyed to hear that our application was successful, a milestone for any professor. My first NSF grant! I quickly made plans to attend the International Ecological Congress, a large conference held that August in Yokohama, Japan. Afterward, Nakano and I could visit the field site in Hokkaido where he had worked the previous year, to conduct pilot work for our research the next two summers.

This awakening to Japan changed the rest of my career, and indeed, my life experience. But like much that occurs in science, and in life itself, it was an accident, serendipity, a paradox. I traveled to Japan expecting to learn relatively little beyond seeing a new country. Instead, I found a vibrant, unconventional young biologist whose work was amazing, and yet based on a very different culture and tradition of study from my own. Our initial contact developed into a close friendship, built on an intense passion to understand why charr in northern Japan occur where they do and behave as they do. This work enriched more than a decade of my career. And through our collaboration Nakano also developed his own skill and stature, which allowed him to expand in an entirely new direction, crossing more boundaries to discover a new understanding of why streams are important to their surrounding forests. But it also led to tragedy, and loss, and finding ways to renew hope and transcend that loss. We followed clues that Nakano left us about the next frontiers in science, about man's influence on streams and forests. Ultimately, the results of this further work led me to ask about the very future for streams and rivers, and why sustaining them in our future could be essential for us as humans.

Chapter 2
Exploring the Mystery of Charrs

The trail grew steeper toward the headwaters along Poroshiri Stream, tumbling cold over boulders and cobbles past remnants of winter snow on its way to the Niikappu River. The stream seemed to spring directly from the thickly forested mountain valley above, and the valley seemed suspended from the lowering misty clouds that shrouded Poroshiri Mountain. As I hiked, the trail sought passage among large elms and maples, and scattered spruce and fir, looking very similar to trees I knew from North America. But the low plants covering the ground were odd and unfamiliar, with tough shiny leaves that appeared almost artificial. This *sasa*, a low-growing form of bamboo, reminded me that I was in northern Japan.

Shigeru Nakano and I had traveled long to reach this place, among the most remote parts of northern Japan. The trip had taken five hours of driving from Tomakomai Experimental Forest, a research station of Hokkaido University located near the sea coast southeast of Sapporo, the capital city. For the last three hours we had traversed dirt roads that wound along the shores of a large reservoir filled with aquamarine blue water, through three tall locked gates of formidable steel bars, and finally along narrow and steep mountain roads carved into the sheer cliffs above the river. Beyond the rustic A-frame cabin that became our base camp were miles of wilderness reached only by trails, which led deep into the Hidaka Mountains that form part of the north-south spine in the middle of Hokkaido Island. Here, native charr and brown bears live landlocked in headwater streams and forests, beyond the last reservoir and any human habitation.

The exhilaration that flooded my mind and body washed away the traces of jet lag from my extended journey. I was finally *here*, and finally doing this work on native charr that we had planned for so long. On this June morning in 1991 I set out to reach the thermometers I had immersed several places in the stream in protective plastic pipes only the week before, to read and record the maximum and minimum water temperatures. But I was also keenly aware that I was hiking alone. All my senses were heightened to make sure that I didn't slip stepping over logs or between boulders and risk injury. My eyes probed the landscape ahead for any signs of animals or other dangers. I blew my whistle frequently as I climbed higher along the trail, hoping to alert any bears within earshot.

Although I've often seen bear footprints and bear scat, I have never had a close encounter with a bear while doing field research in Japan, North America, or anywhere else I've worked. But many of my Japanese colleagues seemed very frightened by the possibility. Tetsuo Tanaka, Nakano's collaborator on their study of charr feeding behavior in this same watershed in 1989, had gestured excitedly about the *higuma* (bears) in these mountains when I met him at the conference in Yokohama in 1990. But Japan has no other large mammal predators such as mountain lions or wolves, which occur in the places I've worked in western North America. So I thought, "How dangerous can these bears be? They are probably small, like the black bears I had seen in Minnesota and the West." Not long after, I learned I was wrong.

After the first ten days of field research in 1991, we returned to Tomakomai and visited the main campus of Hokkaido University in Sapporo. Nakano introduced me to Dr. Aoi (pronounced *aa-oh-ee*), one of the experts on bears in Hokkaido. "So, what is this bear that you have in Hokkaido?" I asked. "Oh, it's the same species as your brown bear in Alaska," he replied. "What?!" I exclaimed, in great surprise. Apparently, brown bears first emigrated from Asia, where they evolved, to North America across the Bering land bridge only about 50,000 to 70,000 years ago. Sea levels dropped throughout the world when glaciers advanced during the Pleistocene Epoch because more water was frozen in the ice, so the shallow sea between Siberia and Alaska became land. As a result of these relatively recent connections, Alaskan brown bears, and the grizzly bears that evolved from them, are very close ancestors of the Asian brown bears in Kamchatka and Hokkaido. After learning this, I realized I needed to be even more careful, and kept the bear mace and my whistle close at hand on every trip. But, I thought, "What will an ambling brown bear do when I stand up from a pool, after lying completely still for so long encased in a black nylon dry suit? Would they ever see or smell me?" I just hoped that I didn't need to find out the answers, and decided to make lots of noise when hiking and emerging from streams.

Why had we come to this remote location in the first place, far from our homes and families, to risk encounters with bears and our personal safety working alone in the mountains of Hokkaido? Planning our study had taken two years, writing letters and sending fax messages back and forth across the Pacific Ocean. We refined our ideas, wrote a detailed proposal to get the funding, and even traveled to this stream after the Yokohama meeting the previous year so I could visit the field site and snorkel to see the native charr we would work with. Our passion was to follow the clues that Dr. Ishigaki had published in his little yellow book and to cross new boundaries in exploring the mystery of charrs in northern Japan.

But we also wanted more. We wanted to help answer a big and important question that is at the core of the entire field of ecology. Why do species occur where they do? And what about species that are very similar, like Dolly Varden and whitespotted charr? How do they divide up the habitat in a watershed, and how do they coexist where they occur together in the same habitat? Ecologists hope that if we can know these things, then we can predict where these animals will disappear, or become more common, as humans change the habitat, and the Earth's climate, by their actions. Most ecologists are "on a mission" to seek answers to these kinds of questions.

"OK . . . now we will speak Japanese," announced Nakano enthusiastically, driving away from the parking lot at the Sapporo airport where I had just arrived. My bags and backpacks containing dry suit and field gear were safely stowed in his SUV, which we would use for our expedition into the mountains. I realized only much later that I had suggested in an early letter that I would need to learn to speak some Japanese. Now I apologized to him that I would not be able to do it "cold turkey." Since he could speak some English, but I could speak no Japanese, it didn't seem practical for us to attempt rigorous and complicated field research in a language entirely foreign to one of us. Fortunately, I found out the next day that Nakano had arranged for a rather tall and reserved graduate student, Satoshi Kitano, to join our expedition. Kitano spoke English quite well, often using fairly elegant words. He became our "walking dictionary."

But I had much to learn about Japanese culture, custom, and food, and encountered many new vistas in the Hokkaido landscape during this first summer. First, the staple foods that most field biologists in North America eat—such as peanut butter, bread, and cheese—were virtually absent from the supermarket in Tomakomai. Huge bags of rice filled an entire aisle, and noodles another. Fishmongers yelled loudly to shoppers, directing them to the long cases of crushed ice with the largest assortment of the freshest fish I had ever seen. Odd-looking vegetables were interspersed with familiar ones, all very expensive. But peanut "cream" came in jars that held only about three-quarters of a cup (and no chunky variety), and all bread was soft and white and sliced an inch thick. The only cheese available was also soft, white, and came only in small packages. I enjoy many new foods, but wondered how I would find enough to eat during long days of snorkeling in cold water. Fortunately, there was plenty of rice and noodles, and Nakano was an excellent chef in the field. His parents had owned a restaurant, and because everyone else was usually busy preparing dinner for guests, he had learned how to cook for himself. Perfectly prepared fish and other meats, vegetables, and fluffy rice appeared in short order each evening from two single-burner camp stoves with Nakano and Kitano at the helm. Although I am also a good cook in camp, I have never eaten such well-prepared and delicious meals in the field.

Waves dashed the rocky sea coast of southern Hokkaido, sending spray high in the air as we set out on the long drive for the first ten-day stint of our research expedition. Nakano's vehicle was completely full with all our field gear and boxes of fresh, dry, and canned foods. We turned north at the mouth of the Niikappu River, where it flows into the cold North Pacific Ocean, and entered a world of rice fields, cattle pastures, and horse ranching in rural Hokkaido. Everything was smaller compared with farms I knew from the midwestern United States. The fields and pastures, the barns and houses, even the tractors and hay baling machines, were about half the size. They were made to fit the smaller plots of land that people had carved out of the river valleys and mountainous terrain of northern Japan, a land shaped by volcanoes and earthquakes in the Ring of Fire. Streams were channelized and straightened to make room for rice paddies and hay fields, and even the sloping sides of the valleys had been cut away to create more bits of flat land. Nature had been tamed to serve man in the lower Niikappu River valley.

We turned onto a dirt road and soon encountered the first large iron gate, and then another at the first reservoir. Nakano opened the first gate with a key, and produced a bottle of whiskey to give the reservoir gatekeeper to gain passage through the second one, joking to smooth the transaction. I learned to count in Japanese (*ichi, ni, san, she, go, roku . . .*) as we wound along the shoreline of the reservoir arms, deeper into the mountains. I felt like a child in a country where I didn't understand the language, or much about the food or customs, and so couldn't yet travel by myself. But it was exciting to learn the new language and landscape and experience new vistas. Beyond the last locked gate the grade rose sharply, and soon we were climbing on a narrow road hung along sheer cliffs, driving much faster than I wanted to go. I had traveled many narrow mountain roads to field sites in Colorado and knew what precautions were necessary to be safe. I didn't want to die at the bottom of a remote mountain canyon in Hokkaido before our work even began. "Nakano—what is the word for 'slowly'?" I asked. "*Yuukuri*, Fausch-san," he replied.

What *does* drive the distributions of different species on the landscape, this fundamental question for all of ecology? Is it simply an accident of history that one species ends up only here and another over there? Does one fish gain access to headwater streams because it arrives first, whereas another is blocked by a waterfall that forms later and so becomes restricted to only downstream reaches? Or, do the two species simply tolerate different conditions, such as colder versus warmer temperatures, and so become separated into different habitats? Or, does one species drive the other out, relegating the subordinate to only those places where the dominant species can't live? Ecologists came up with this last idea because they noticed that some species extend their range into more habitats where historical accidents like waterfalls or mountain ranges excluded their competitors.

The challenge for ecologists and evolutionary biologists is not so much to come up with these different scenarios to explain species distributions (or "hypotheses," in the language of science). We train ourselves and our students to be creative and imagine different likely reasons that could explain the facts. The greater challenge is to decide what key pieces of evidence must be gathered in the real world to prove which one of these explanations, or which combination of them, is closest to the truth.

Adding to this challenge is that each explanation depends on processes that operate at a different scale in time and space. So, often it requires completely different methods to gather the evidence to test the different explanations. For example, understanding the role of historical accidents requires delving into the geological history of large regions played out over millions of years. In contrast, determining which species wins in competition requires careful study of individual fish in specific pools in the "here and now." And, it could be that competitive ability itself changes with environmental conditions such as temperature. It makes sense that one species may be dominant at warmer temperatures, whereas the other could be dominant at colder ones, a phenomenon ecologists call "condition-specific" competition.

More than 800 years ago, on April 25, 1185, *samurai* warriors of the *Genji* (white) clan won a final decisive battle against the *Heike* (red) clan at *Dan-no-ura*, a narrow strait at the southwest tip of Honshu, the main island of Japan. The history of this sea battle is celebrated in plays and books and is known to every Japanese person. It ended a long period of relative peace, after which Japan descended into more than 700 years of feudal rule, the first *shogun* being a leader of the victorious *Genji*. In his book of beautiful Japanese prose, Dr. Ishigaki used this legend of ancient conflict in his final chapter as a metaphor to explain how whitespotted charr (*iwana*) and Dolly Varden (*oshorokoma*) came to be distributed the way they are in Hokkaido. The white clan had invaded, and overpowered the red one.

For his PhD research, Ishigaki had sampled charr from many streams throughout Hokkaido and had noticed several things. First, where the two species occurred in the same river basin the Dolly Varden were always upstream and whitespotted charr always downstream, and they typically overlapped only in short reaches at the point of contact in any

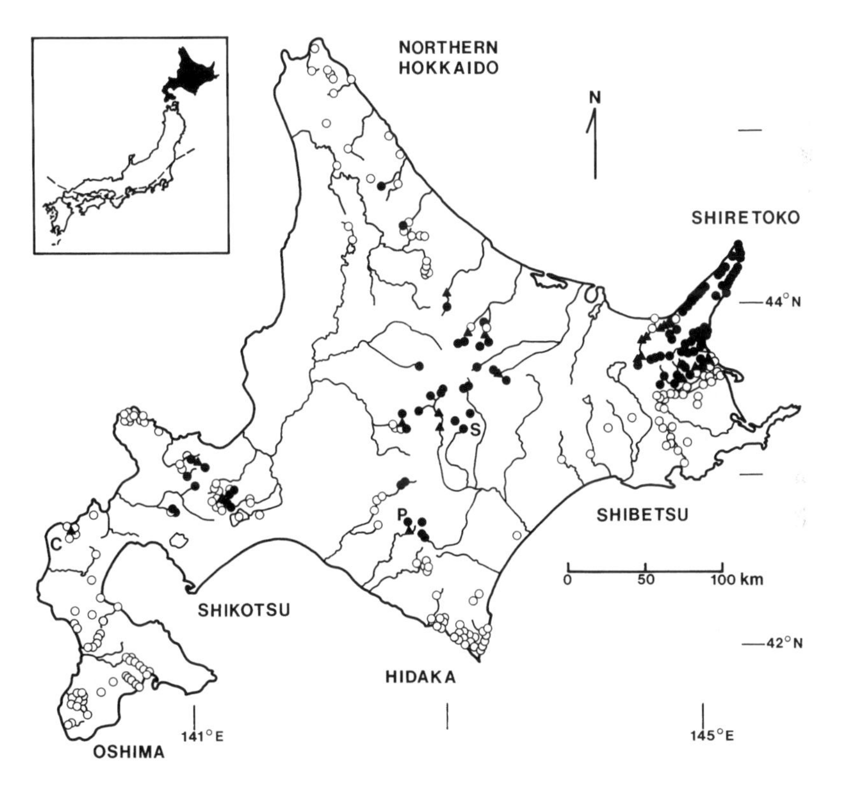

Figure 4. Locations where Dolly Varden (filled circles) and whitespotted charr (open circles) were captured by Ishigaki and other fish biologists across Hokkaido Island. Triangles show the short zones of overlap where both species were found together in streams, such as in Poroshiri Stream (P). In Japan, Dolly Varden occur only in Hokkaido, whereas whitespotted charr range to the southern end of Honshu (dashed line in inset). The location of Lake Shikaribetsu (S) is also shown. Names refer to regions across Hokkaido. Reprinted from Fausch et al. (1994), with permission.

given stream (see figure 4). Second, their distributions shifted from the ocean toward the headwaters as the climate changed from east to west across the island (see figure 5). For example, in streams draining the cold Shiretoko Peninsula at the northeast tip of Hokkaido, Dolly Varden extended all the way down to the ocean. No whitespotted charr occupied these steep cold streams, even though seagoing ("anadromous") fish of this species (called *amemasu*, meaning "rain trout") have been captured all around Hokkaido Island and could easily colonize these streams. In contrast, on Oshima Peninsula, at the farthest southwest tip on the opposite end of Hokkaido where the climate is warmer, whitespotted charr

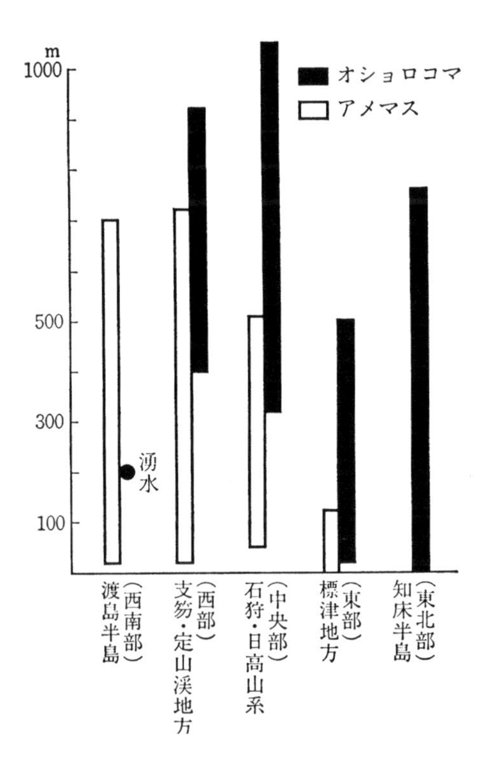

Figure 5. Figure from Ishigaki's book (1984; reprinted with permission) showing the distribution of Dolly Varden (black bars) and whitespotted charr (white bars) with elevation (in meters) in five regions across Hokkaido (see figure 4). From left to right (west to east): Oshima Peninsula, Shikotsu, Hidaka Mountains, Shibetsu, Shiretoko Peninsula. In Oshima Peninsula, Dolly Varden occurred only in one cold spring, shown as a filled dot.

penetrated to the headwaters. The few Dolly Varden populations found there were landlocked above waterfalls or inhabited cold spring streams.

These charr distribution patterns led Ishigaki to suggest that Dolly Varden had colonized the island first during an earlier and colder period. However, as the climate warmed after the most recent ice age, whitespotted charr invaded the island and pushed the Dolly Varden out of the warmer habitat in downstream reaches, especially in southwest Hokkaido. Ishigaki thought that whitespotted charr were better competitors in the warmer reaches, but also speculated that Dolly Varden could still outcompete whitespotted charr in the coldest and steepest streams. As a result, Dolly Varden found refuge only in the headwaters of mountain streams like Poroshiri, and the "last palace" of Shiretoko Peninsula, the remote peninsula of steep terrain jutting out into the cold North Pacific Ocean. No whitespotted charr had penetrated streams in this last bastion.

One of the graduate students who accompanied Nakano and me to Poroshiri Stream to conduct pilot work in August 1990 is now one of the leading experts on the evolution of charrs in Japan. Shoichiro Yamamoto and his colleagues have spent much of the last fifteen years using new genetic techniques to trace how these different charr species are related and provide clues about where they came from. The current distributions of these fish show that whitespotted charr occur farther south and Dolly Varden farther north. Whitespotted charr range from the mountains near Kyoto on the main island of Honshu to the northern end of the Kamchatka Peninsula in far eastern Russia. In contrast, Dolly Varden range from the southern tip of Hokkaido Island across the Bering Strait to Alaska, and south to Washington state. However, each species is made up of many different "lineages" in their family tree, and the distributions of these fishes have ebbed and flowed over millions of years.

The best evidence to date suggests that a former ancestor to both charrs evolved on the mainland in far eastern Russia or in the Japanese archipelago. Genetic data indicate that the early prototypes of the two species diverged 3 million years ago, and that the lineage of Dolly Varden that ended up in Hokkaido diverged from other forms of this species about 1.5–2 million years ago in watersheds around the Sea of Japan, which lies between Japan and the mainland. Evolution of different species and lineages often requires some form of isolation, which allows small initial differences to evolve into larger ones. When sea levels dropped during the repeated periods when glaciers advanced, Japan became attached to the mainland and the Sea of Japan became a nearly enclosed body of water. This gulf and the rivers flowing into it would have provided the isolation needed for a new charr to evolve. As sea levels rose again, anadromous fish like these charr could move freely through the ocean and colonize new islands. These cycles may have been repeated several times. In the end, Ishigaki's hypothesis remains plausible to explain current charr distributions in Hokkaido, even thirty years later. But what is the role of things that affect charr today, like temperature and competition?

We wanted to peel away another layer of the onion. As scientists, we often think of our questions as a big onion, because although each study reveals one more layer of the truth we seek, we realize that there are many more layers beneath. No one study is comprehensive enough to provide a complete answer to an important question in science. Ishigaki had revealed the outer layer with his pioneering work on these charr, but Nakano and I wanted to go deeper. We wanted to understand how important temperature really is in explaining the shift in charr distributions across Hokkaido Island, and which species has the competitive advantage over the other where they occur together.

We thought that if temperature was the main factor explaining why the charr shifted across the island, then the boundary between the two species along streams should occur at the same temperature, regardless of the region and elevation. However, we also reasoned that if their distributions were driven *only* by temperature, and not by any competitive interactions, then the species should not care about each other and never interact. But we already knew from snorkeling observations in Poroshiri Stream by Nakano and Tanaka, and from earlier experiments in aquaria by Ishigaki, that this was not true. The two charr species *were* aggressive toward each other and understood each other's language of behavioral displays when they competed for profitable feeding positions in pools during summer. So, if Ishigaki's hypothesis that whitespotted charr had driven Dolly Varden from downstream reaches was true, then we predicted that whitespotted charr should be the dominant species in the narrow zones of overlap at the boundary between them. Ultimately, we also wanted to know whether temperature could change this competition, so that Dolly Varden could defend their "last palace" in the colder and steeper reaches upstream, and in Shiretoko Peninsula.

It had become my dream to rekindle research on whether competition between trout species could explain their distributions in streams, the topic I first studied in graduate school. For many scientists, their first topic remains their first love. However, there are many different facets to this interaction between two species, and competition is incredibly difficult to prove convincingly, especially in the field. A key question is whether in the zone of overlap the two species treat each other essentially as the same species, so that individuals of the same size are equally matched. In contrast, it could be that smaller individuals of one species can win over slightly larger individuals of the other, creating a pecking order that depends not only on size, but also on the species. After much trial and error, we designed a field experiment that could test this idea.

"We must not fail," emphasized Nakano quietly, as we discussed our plans while standing outside the A-frame cabin one day early in the first summer. I felt exactly this same weight of responsibility. The funding that each of us had received was very difficult to get, and we both knew that this was our big break at this point early in our scientific careers. And yet, like much of the research that I've done since then, our first attempts did fail. I had planned to conduct the experiments on competition between the charr in pools, but the flow was so high from melting snow in June 1991 that only a few were suitable. Most were simply too swift to be considered pools at that flow, and were drowned out by the high water.

Originally, I also planned to remove all fish from the pools and stock specific numbers and sizes of each species for the controlled experiment, but new fish kept immigrating daily from somewhere else. Day after long day we worked hard, removing fish, marking and restocking the ones we had selected, and snorkeling to measure the results, but met only failure. In addition, Nakano wanted to do a different study than I did, to continue his work on the feeding behavior of charr and how it was affected by the amount of drifting food. We argued, and avoided each other for a day by working on our separate projects. But soon after we reconciled, agreed to join forces to help each other, and planned a grueling field schedule. From these chaotic, disorganized, and disheartening beginnings grew research that helped reveal deeper truths about these beautiful charr.

Fishing underwater. It has to be the coolest experience a fish biologist could imagine. Like most, angling was a passion for me growing up, and I became fascinated by watching fish respond to live prey used as bait. I had settled on the idea that the easiest way to measure which species was dominant over the other was to remove the most dominant charr from the pecking order in pools, one at a time. Nakano had perfected his method of using traditional Japanese *tenkara*-style fishing to capture the dominant charr while snorkeling. He simply drifted a live stonefly or caddisfly larva past their position at the head of the pool, and set the tiny hook quickly when he saw the fish rise to capture it. We never killed or injured any fish caught this way.

After the dominant fish was removed, each subordinate charr shifted in turn by one notch to a better position. In combination with our data on which fish won the contests for these positions, this allowed us to quickly measure the entire dominance hierarchy between the two species. Ultimately, our results showed that these charr were nearly identical in their competitive ability in the zone of overlap, and that dominance was primarily related to their size alone. Overall, they treated each other as the same species. But why didn't this match our prediction that whitespotted charr would be the dominant species where they occur together? Only later would this puzzling result make sense.

I was never as good at the diving as either Shigeru Nakano or Satoshi Kitano. The water was frigid, averaging only 45°F to 48°F, and my thin frame could not maintain body heat for long at this temperature, even in a dry suit with several layers of fleece clothing. After only forty-five minutes of lying completely still to avoid disturbing the fish, my mind wouldn't maintain focus on recording the data, and I knew I needed to get out and warm up. I'd feel myself slowly becoming hypothermic. I'd clamber stiffly out of the cold water and drink hot tea, eat chocolate candy, and do jumping jacks on the banks of the stream, all to regain physical and mental stamina for the next bout of observations. Even so, writing with cold hands in awkward cramped positions underwater made my data sheets difficult to read. Each evening, I spent many hours after dinner around the wood stove in the cabin, erasing and rewriting the data for each fish. By contrast, Nakano's data sheets looked like he had written them while sitting comfortably, warm and dry, at a desk.

Cherished memories of my family often penetrated the fatigue during the rigors of fieldwork and have always carried me through the difficult times. Although my Japanese colleagues always treated me like royalty, I began to miss the familiar things of home, and the chance to speak "fast English" to someone. But a song from Ben's kindergarten music program a few weeks before about a boy and a bear never left my head while snorkeling in Hokkaido. Since he was a little boy, Ben has always been a creative musician and continues to perform while pursuing a career where he can use this creativity. And from his older sister Emily, all of us learned the value of persistence. Emily has disabilities, but she never let that slow her down in pursuing what life has to offer. Although a nurse told my wife Debbie when Emily was a baby that she would never walk, talk, or go to school, we stubbornly rejected that prophesy. We decided to bombard her with every stimulus and opportunity possible to help her overcome her delays. When she met barriers, she simply chose another door. A few years ago, she earned her Master of Arts in Special Education. Persistence and determination alone are omnipotent. And so how could cold water or exhaustion from long days working in the rain prevent me from reaching our goal of understanding the mystery of Hokkaido charr? Sweet Emily, then eight years old, was a source of inspiration.

Nakano, too, shared this sense of opportunity and passion. His older brother, Minoru, was stricken with multiple sclerosis and was not able to be active as an adolescent or adult. Like Ben, he is a creative musician, and like Emily, he has achieved great things by composing music despite his disabilities. During long discussions on the cold evenings around the warmth of the wood stove, Shigeru and I realized we had experienced some of the same bittersweet longings that come to rest between loss and humble gratitude. Our sibling and child had suffered, and yet our families had found ways to help them triumph despite these disabilities. And in these triumphs, we enjoyed an even greater depth and fullness of life than we would have otherwise, drawing our families even closer. In the end, I believe the same question formed in each of our minds, and inner voices spoke the same thought. Given the difficulties that others faced, how could we make the most of the chances we had been given? Life and health are precious, and we both felt driven to make the most of these gifts.

Cold gray light filters among the trees and steals gently through windows of the A-frame cabin only four hours after midnight, softening the outlines of the wood walls and timbers. There is no daylight saving time in Japan to alter human activities. Shigeru Nakano stirs from his sleeping bag laid on the rough wood floor, groans and stretches to ease sore muscles, and sets the teakettle on the camp stove to boil. Breakfast is made and eaten while assembling equipment and data sheets for the new day and gathering dry suits hung to dry. Satoshi Kitano emerges through the creaking wooden front door, raincoat and waders dripping from the steady rain outside but happy to have found a few hours at dawn to go fishing. It is another cold and rainy day to endure.

After breakfast, Nakano strides off quickly for two hours of snorkeling to make underwater observations at Pool A, a hundred yards downstream, and then takes a quick break at the cabin for hot tea to warm up and a chance to find a thicker pair of diving gloves. Then he makes another bout of observations at Pool B several hundred yards farther downstream, while we others capture charr in an upstream reach with a cast net to gather needed data on their lengths and ages. A late lunch of steaming ramen noodles, a short rest, and then Nakano snorkels a final bout of nearly three cold hours underwater again at Pool A. The rain never stops, it seems, blurring the boundary between stream and sky. Late in the day, the fish collected are measured in the refuge of the cabin and preserved and safely stored. Dinner brings satisfaction to hunger from the energy given over to cold water and rain, but washing dishes sharpens aching in the shoulders. The plastic data sheets are gathered from their ranks drying on the wooden floor and pored over beneath flickering gas lanterns in the warmth from the stove. Work finished, the data sheets are organized and stored safely in shallow boxes for each pool. Brushing teeth outside in the cold drizzle awakens exhausted senses, but merciful sleep comes quickly when bodies curl into sleeping bags again.

About four of our allotted five weeks of field research had passed during that first summer in 1991, but I wasn't sure if we were succeeding or not. We focused on three questions the first year, working every waking hour for ten-day stretches, through rainy days as well as sun. First, I wanted to know whether the limits of the two charr species in Poroshiri Stream and its main tributary, *Migi Sawa* (Right-Hand Stream), occurred at the same boundaries of temperature, stream slope, or some other physical feature that could explain them. For example, did whitespotted charr drop out at the same altitude in both streams, and was the water temperature at these locations also the same during summer? We spent many days collecting fish by angling, cast nets, and electrofishing, and measuring these physical factors at many points along each stream to test this idea. Electrofishing doesn't kill fish, but uses direct current (DC) electricity to force fish to swim from their hiding places toward a handheld electrode where they remain stunned until netted. However, I soon learned that Nakano's ability to catch fish by *tenkara* angling (even without snorkeling) and Kitano's skill with a cast net rivaled the backpack electrofishing gear that I knew best and originally thought would be most effective.

Second, we continued testing my ideas about competition between the two charr species by removing individual fish from dominance hierarchies in three different pools. From the many hours of snorkeling to record the positions of the remaining fish we amassed a sheaf of the thin plastic data sheets about a half-inch thick from each pool. And third, we continued diving throughout nearly a mile-long section of Poroshiri Stream to test Nakano's ideas about what explained the feeding strategies of the two species. For example, his earlier research with Tanaka in August 1989 had shown that the majority of Dolly Varden charr were foraging widely over the bottom and picking bottom-dwelling insects. In contrast, in June 1991 we found that most were holding relatively fixed positions in the current and capturing drifting insects. Could it be that as the remaining snow melted and the stream flow subsided, drifting insects became scarce and Dolly Varden were forced to shift their feeding? Nakano thought that by recording this fish feeding behavior at several times during summer as the flow subsided and also sampling drifting versus bottom-dwelling insects, we could compare the data and test this idea.

But one day I stumbled across another way to test Nakano's idea. As I described above, scientists are always thinking about what critical piece of evidence can convincingly prove or disprove a hypothesis. I had lain down in a small side-channel of Poroshiri Stream, simply to test whether my mask was watertight, before I entered a pool to record fish observations. Clearing water from your mask can disturb fish and ruin an observation session, so it pays to check beforehand. A single medium-sized Dolly Varden held a position a few meters upstream from me and was making short rises to catch drifting invertebrates. "What if I simply filtered out that drift farther upstream?" I thought. "Then we could test whether declining drift actually causes the shift in feeding behavior." After all, it might be that the fish shifted because the stream was gradually warming throughout the summer, or because the days were getting longer and the fish simply became more active. Or perhaps there was some other reason beyond our ken.

I realized that what we *really* needed was a field experiment to cut off the drift. So I ran back to the cabin, found one drift net, and carefully anchored it several meters upstream from the fish to filter out the drifting insects. I waited a few minutes for the fish to settle down after the disturbance and again took up my position downstream. Without the normal supply of drifting insects the charr shifted within a few minutes to ranging widely and picking insects from the streambed, the same behavior we had seen by the smaller Dolly Varden in our experimental pools. This kind of intuition and serendipity, coupled with trial and error, are often the raw materials for scientific discoveries. We resolved to do these experiments in entire pools the next year, which I described in the previous chapter.

So, in the end, were we able to unravel the mystery of charr in Hokkaido? Did we manage to peel away more layers from this big onion? The first bit of truth we discovered was that no one factor can explain everything about why the two charr occur where they do in Hokkaido. There is no silver bullet. During their evolution, both charr arrived at the island, and their distributions may have ebbed and flowed through the millennia. As Ishigaki proposed, it seems likely that Dolly Varden arrived first, because it occurs above some waterfalls that excluded whitespotted charr. But a more detailed explanation may be offered when more work is done by evolutionary biologists on this question. Modern genetic techniques may allow them to reach back into the past ancestry of these fish, cloaked in the shadowy history of many millennia, and peel away more layers of that onion.

As for the factors affecting charr now, when we considered the whole island and each of the large drainage basins, we found that water temperature is apparently a main driving force. The evidence for this is that the narrow zones where the two species occur together are very similar in temperature, regardless of their elevation or region. As described above, this zone shifts from lower elevation in northeast Hokkaido where it is colder, to higher elevation toward the southwest where it is substantially warmer. In contrast, no other variable, such as stream slope, explains charr distributions at this broad spatial scale. However, local factors, such as shallower pools and stronger floods in *Migi Sawa*, appeared to account for why the upstream boundary for whitespotted charr was at an elevation about 100 yards higher in Poroshiri Stream than its neighboring tributary. The harsher habitat in *Migi Sawa* apparently prevented whitespotted charr from living farther upstream. Local factors make local adjustments to the broadscale pattern.

Although we originally thought that Dolly Varden would lose in competition to whitespotted charr in the zone of overlap, to our surprise we instead found them to be equal competitors. The two species treat each other as nearly identical, so that the dominance hierarchy that develops is based on the size of individual fish rather than their species. Large fish can win better positions over smaller ones, regardless of species. However, during the "crunch" period of late summer, when drifting invertebrates become scarce, smaller Dolly Varden can also quickly shift to feeding on the bottom-dwelling insects and avoid the competitive pressure from larger charr for prime feeding positions. We believe this allows Dolly Varden to persist through these lean periods and better coexist with whitespotted charr in this zone.

After two summers of grueling research and several more years of analyzing field samples and the data themselves, and writing manuscripts, we felt like we had peeled back several layers of the onion. However, we had not answered Ishigaki's main question about whether Dolly Varden can win at colder temperatures and whitespotted charr at warmer ones. This would require another quirk of fate, and meeting another young Japanese student who wanted to become a scientist and explore the mystery of charrs further.

‹ ›

I noticed his jet-black hair and Asian features among the sea of more than a hundred faces in my junior-level course in fish biology (Ichthyology) at Colorado State University. He was clearly struggling to understand my lectures. I purposely slowed down a bit and tried to remove all the slang from my speech, just like when I gave talks in Japan. Young Yoshinori Taniguchi had followed his sister to CSU and planned to earn a second degree in fishery biology. His father raised *koi* (fancy goldfish) in Japan, and Yoshi, as he calls himself, had originally studied aquaculture. Afterward, he decided that he wanted to break from tradition and work with wild fish, rather than farming his father's *koi*.

"I am working in Japan right now, on charr in Hokkaido," I said, approaching him in the laboratory portion of the course where students learn the practical skill of identifying fishes. His eyes grew wide in amazement. When we talked after the class, I learned that he had just read Ishigaki's little yellow book while traveling on the plane to America. More serendipity. Later, I hired Yoshi to measure the stream slope at the 360 different locations where Ishigaki and others had collected charr, from detailed maps of streams in Hokkaido. I'm not sure who else I could have found to read the names in Japanese characters on the maps. After he took my senior-level course in fish ecology and completed his undergraduate degree, I recommended Yoshi for an opening as a graduate teaching assistant at the University of Wyoming, which allowed him to pursue his master's degree. Based on his excellent work there, I was then able to introduce him to Shigeru Nakano in Japan. Nakano accepted him into the PhD program at Hokkaido University, and they began work at the Tomakomai Experimental Forest research station where Nakano had become director in 1996 after Dr. Ishigaki retired. Yoshi's dream of studying Hokkaido charr was about to come true.

Yoshi and Nakano realized that if Ishigaki's idea was correct, then they should be able to measure in the laboratory whether changes in temperature altered which charr won in competition—the idea of condition-specific competition. And, because young fish are more sensitive to this aggressive competition than older ones, they decided to measure how it affected the foraging, growth, and survival of these charr during their first year of life. They constructed twelve artificial streams for the research and compared each species alone, and then both together, at cold versus warmer temperatures. They chose temperatures similar to those found above and below the zone of contact in Hokkaido streams, based on our previous data.

In the end, the results complemented our field research very well and helped explain another layer of truth about the distribution of charrs in Hokkaido. Indeed, whitespotted charr outcompeted Dolly Varden at warmer temperatures such as those in downstream reaches of Hokkaido streams where only whitespotted charr occur (54°F). They excluded Dolly Varden from the best foraging positions and grew faster and survived better. Therefore, the downstream limit of Dolly Varden distributions in Hokkaido streams is very likely determined by the point at which they lose in competition to whitespotted charr as the stream warms.

Surprisingly, however, the upstream limit of whitespotted charr is apparently *not* set by where they lose in competition to Dolly Varden. At 43°F, a typical summer temperature in upstream reaches of Hokkaido streams where only Dolly Varden occur, the two were equal competitors for feeding positions, and whitespotted charr even grew faster than Dolly Varden. However, their critical finding was that far fewer whitespotted charr survived at this temperature, which is apparently too cold for them to tolerate, with or without Dolly Varden. They simply cannot persist at these cold temperatures during their first year of life, which then sets the upper limit to their distribution, regardless of the presence of Dolly Varden.

But why did we find that the two species were equal competitors in the zone of overlap where they occur together in Poroshiri Stream? This relatively short stream segment, only about 1.2 miles long, is similar to many other short zones where the two species coexist in Hokkaido watersheds. Why didn't whitespotted charr win, and exclude Dolly Varden from this place too? Two reasons seem clear to me now, twenty years later. First, water temperatures during summer averaged 46°F–48°F in this zone, right in the middle between the two tested in the lab, which may be a tipping point for their interaction. Second, Dolly Varden can probably coexist with whitespotted charr only where the two have nearly equal survival. One thing that helps "even the score" is the ability of Dolly Varden to shift feeding behavior during crunch periods in late summer when food is scarce. In contrast, farther downstream where the balance of competition tips, Dolly Varden lose and gradually die out, which creates their downstream boundary. Upstream, whitespotted charr die out because it is too cold. Together, Yoshi and Shigeru and I had come close to a central truth, beneath many layers of the onion.

More than a decade passed between when I first wrote to Hiroya Kawanabe asking for his help to work with Nakano and when Shigeru and I and Yoshi had finally completed all our work and published it for the rest of the world to see. Nakano and I had crossed many boundaries—of our different languages, cultures, and ways of doing science. We developed a deep and rewarding friendship, forged from many long days in the field working together and depending on each other for critical support and safety. Our bond was strengthened during even more long months painstakingly analyzing data and writing manuscripts, sending messages with them back and forth across the Pacific Ocean by fax, e-mail, and letter. Scientists count themselves lucky if they find even one collaboration so close as this throughout a lifetime of work.

Nakano found funds for me to visit Japan in 1994, and my young family traveled together with his wife Hiromi and their young daughter Nana to see rivers and valleys in his home region of Gifu Prefecture in central Japan. He visited me in the United States in 1995, and I came again to Japan in 1998 to do research with Yoshi. During all of these travels, I saw many beautiful riverscapes throughout Japan. Many looked like the paintings on old Japanese scrolls, of steep-sided mountain valleys and rivers shrouded in mist, with gnarled trees protruding from cliff faces and forested mountains above.

But I also saw much destruction of Japan's beautiful landscapes, rivers, and charr and salmon. The long trips across Hokkaido to reach Poroshiri Stream passed many channelized rivers and streams, some entirely paved with cement blocks to prevent any erosion of their degraded channels. Salmon were blocked from most by weirs or dams. I learned from Nakano that remote streams like Poroshiri were the only places he could study charr, because there were almost no angling restrictions in Hokkaido. Anglers depleted charr from most streams, including those in national parks. Even in Poroshiri, town leaders and their friends would come a few times each summer and fill small coolers with fifty to a hundred charr, apparently thinking the stream could support such harvest. And yet, the plight of our native cutthroat trout in Colorado and many places throughout the western United States is not so different. Many streams are channelized, or the habitat is destroyed or simply dried up to supply water for cities or farms. The remaining native trout populations occur only in small remote streams deep in the mountains, like Poroshiri.

In many areas of my work, I began to ponder the future for these beautiful fish and the landscapes and riverscapes that support them, around the world. Both Shigeru Nakano and I began other research to reveal how these fish use habitats throughout networks of streams, and how these streams are linked to their surrounding landscapes. Understanding these linkages led us to a deeper knowledge of how closely connected streams are to the rest of the watershed, and why it is important to keep these connections if we are to conserve not only these fish but many other animals that depend on streams.

Chapter 3

Riverscapes: How Streams Work

I love to walk along streams in early spring. Autumn in temperate latitudes brings a riot of color to most riparian forests, and leaves settling gently down to join noisy bustling riffles or quiet curling eddies. But spring is very different. Like a youth on his or her first date, the riparian forest is uncertain and awkward yet exuberant in the new-found freedom after release from winter's rule.

Early spring along streams brings dazzling light on last year's dead grasses, bent over into clumps by the winter snow, and on leaves matted into the forest floor. Snow and frost are gone, but the trees are just beginning to reveal the first light green tinges on their barren fingers as buds break open. Without the umbrella of summer leaves, the riparian forest floor is warmed in the brilliant sun, like no other time of year until the leaves fall again in autumn. In Michigan, where I first really found streams, the first bright colors to dot the drab forest floor and greet the spring are the dark green leaves and yellow flowers of marsh marigolds. Later, in Japan, I marveled at the lime green fern fiddleheads poking up through the dead brown plant litter, like so many living violins emerging from the ground, seeking the vernal light that will later be absorbed by the maples and oaks leafing out above.

But in the slant light of late afternoon on those first spring days afield, my eyes are drawn from the verdant green plants to movement in the air just over the stream. Like dust motes in a sun shaft, shifting clouds of tiny insects called midges swarm at head height above the water, catching the light and revealing their mating dance. Drawing closer, I see that one midge within the swarm flies upward a few inches and is followed by others, moving up and swirling back down, like dancers in a miniature waltz. In late spring, I will start to notice mayflies of tan and dun, larger than the midges, drifting on the stream surface, and see them struggle off the smooth water and fly weakly toward riparian shrubs. In summer, females of dark brown caddisflies fly rapidly in zigzag courses close to the water surface, as if chasing some invisible prey, looking for a suitable site to lay their eggs. As the final immature stages of these aquatic insects rise to the stream surface to emerge, and drift there in the currents before flying away, they are silhouetted against the sky and many fall prey to vigilant fish holding positions beneath. Then as evening dims to twilight, swallows and later bats sweep and whirl over the stream to catch these aerial prey. But, where do these tiny insects come from? What do they eat? Are they important to streams, and to riparian forests? In fact, these seemingly insignificant insects hold one of the keys to how streams work.

Viewing any terrestrial landscape reminds us that in order to grasp how it works, we need to understand two things. First, we need to understand the *hierarchy* of how features at larger scales affect those at smaller ones. For example, landforms like mountains and valleys shape how and where water moves, how soil is formed, and, in turn, where vegetation grows and animals thrive. Second, we need to know how these components are *connected* across boundaries in three dimensions. For example, trees and other plants take up water from the soil through their roots, store some of it in their stems and leaves, and transpire the rest into the atmosphere. In turn, animals eat plants and other animals, move the water and nutrients they contain across the landscape, and release them elsewhere when they die. Everything is connected. One can no more understand a landscape by focusing only on one part than one can appreciate a landscape painting by viewing a few glimpses through holes in a curtain draped over it. No one part can be understood fully without considering the connections to other parts, to place it in the appropriate context within the whole landscape.

And so it is with "riverscapes," a term we and other scientists have used to highlight similar hierarchies and connections within streams and between streams and their riparian forests and grasslands. Just as for landscapes, understanding riverscapes requires a continuous view of rivers to learn how features at larger spatial scales such as the geology and topography along entire river segments set the context for those at smaller scales like riffles and pools. Here, too, we need to know how plants are connected to the resources they need and use, which often flow between land and water. We want to discover how animals like insects and fish move across these different scales and are connected in webs of relationships between predators and their prey that may cross the land-water boundary. But if we want to start to understand how rivers and streams work, and their linkages across boundaries to riparian forests, we should begin with an understanding of the parts. We first must find out where the water comes from, where the energy to drive the system comes from, and what feeds the animals that live in both the aquatic and terrestrial worlds.

A Physical System

My friend Charley Dewberry taught me a lot about streams when I first opened my eyes to them, starting in graduate school at Michigan State University. And he is still teaching me, from his three decades of work on salmon and trout in rivers of the Oregon Coast Range. We both enrolled in Stream Ecology in fall 1978, a course taught by renowned stream ecologist Ken Cummins. Charley also turned me on to courses in hydrogeology, about where the water comes from, and landforms analysis (also called geomorphology), about the processes that shape the landscape and its rivers. He and other mentors taught me that if I wanted to understand the ecology of fish and other stream organisms, I needed to understand how processes like glaciation and mountain building (tectonics) shaped the watersheds and river channels in which fish evolved in the past, and in which they live in the present. I have taught this basic principle to generations of graduate students seeking the same path.

Just as we humans are products of our backgrounds and experiences, so is each stream a product of its watershed and the geology that shapes it. Where does the water come from? Some comes from the layers of soil just beneath the land surface and is forced into stream channels when rain or melting snow seeps into the ground and pushes out the older water. But other water that creates the flowing stream comes from deep underground, especially in watersheds underlain by thick beds of sand and gravel. These beds, called aquifers, can store the water from rain or snow that percolates down through the soil and creates a water table in the layers beneath. This water can be ancient, having seeped into the aquifer hundreds to thousands of years before. The aquifer acts like a giant tub of sand, filled nearly to the surface with water. And, this water beneath the surface of the land, called groundwater, moves slowly, by gravity, toward the low points in the landscape. It seeps into hollows to create ponds or lakes, and into stream channels to add to the flow.

Small streams join to form larger ones, and then rivers, which carry water to the sea. There the beating sun evaporates water, which forms clouds, and eventually this water falls again as rain or snow on the land, in the great "hydrologic cycle." In the end, sunlight makes the rivers run, just as it runs everything else on our planet.

But Charley and I learned that what happens to the water in between the rain and the sea can mean the difference between a stream in which trout thrive and one that dries up in the dog days of late summer. For example, if the watershed consists mainly of granite or other hard rocks, perhaps covered by only a few inches of soil like on the Canadian Shield in northern Minnesota and Michigan, several things will happen. First, when the snows melt or the rains come the water will move quickly through the shallow soils and drain off the bedrock into the stream channel, creating pulsed floods but leaving little groundwater to supply flow later. Second, because there is little stable input of groundwater, stream water temperatures will be warm in summer and very cold in winter, and will change rapidly during floods, creating stressful conditions for invertebrates and fish. Third, few minerals and nutrients may be dissolved from the hard rocks by the water, so the stream may be unproductive, stunting plant and animal growth. In contrast, if the watershed has deep aquifers of sand and gravel, and especially if there is soft limestone among the rocks, then water will percolate into the ground rather than running off quickly, temperatures will be stable throughout the year, and more nutrients may dissolve. Plants and fish will thrive because the groundwater will create a steady flow with constant temperatures year round, fostering high growth and survival.

Charley, a Michigan farm boy of slight build and long legs (actually, a slightly shorter and smaller version of me), was the first to teach me that you could predict a lot about the character of streams and their fish from the geology and landforms in the watershed. For example, the world-famous trout rivers in Michigan's Lower Peninsula, like the Au Sable River, are created by deep aquifers of sand and gravel, 800 feet thick, left by retreating glaciers. These huge reservoirs store nearly all the water from rain and snow that falls on the land and release it slowly throughout the year, creating extremely stable flow and moderating water temperatures. This is especially important during winter to prevent ice from forming on streambeds, which is harmful to trout and invertebrates. In contrast, rivers in the Upper Peninsula that drain watersheds underlain mainly by hard bedrock have flashy stream flows, with large flood peaks followed by very low flows between rains. Without a stable flow of groundwater, many stream segments can be too icy in the winter and too warm in the summer to support many trout. In fact, two Michigan geologists who were avid trout anglers were able to predict where they should look for good fishing simply based on maps of the geology.

The Benefits of Natural Catastrophes

Just as humans grow and change from their life experiences, so too do rivers and streams bear the legacies of changes in water and land that occur through time. Streams are often affected by natural forces that look catastrophic when they first happen, like floods, fires, and landslides. However, if these "disturbances" are indeed natural, and not triggered by human actions, we usually find out later that they are needed to create favorable habitat for the organisms that evolved in these systems. It stands to reason.

For example, regular floods that at first appear destructive actually dig new pools that fish need during winter. They also winnow silt from the gravel in shallow riffles where many invertebrates live and the fish spawn. Floods topple riparian trees that later provide shelter for fish and invertebrates, and floods also create clean sandbars needed for the seeds of riparian trees and shrubs to sprout. Streams were meant to carry regular floods, and the ecologists who study streams are usually rather happy when they occur because they rejuvenate habitat for many organisms.

Forest and grassland fires along streams also can cause catastrophes in the short term but are usually beneficial when one takes a longer view. Fish can be killed from the heat, and by the ash that washes into streams and robs oxygen from the water. The next rains falling on the bare soil can also trigger landslides in many regions. Whole hillsides can flow rapidly in a muddy mass into the stream, bringing tons of soil, rock, and whole trees into the channel. And yet, like floods, natural fires and landslides have important benefits. For example, fires allow more sunlight to reach the stream, which fuels the growth of plants (mainly algae growing on the streambed rocks). In turn, this increases the growth of the invertebrates that eat algae, and the fish that feed on these invertebrates. In addition, nearly all fires burn watersheds unevenly, rather than uniformly, creating refuges for fish in unburned or lightly burned reaches. These fish can quickly recolonize the burned reaches and produce a new generation of small fish, which often grow faster and larger than in unburned stream reaches because of the increased food supply. Likewise, the natural landslides bring in gravel and large trees, which are needed to create suitable riffles and pools with shelter for invertebrates and fish. Without them, in some regions the streams will eventually wash away most gravel. This creates uniformly shallow "runs" over bedrock, which are poor habitat for most stream invertebrates and fish.

The Stream as an Ecosystem

In 1935, English ecologist Arthur Tansley realized that to understand why plants occur where they do (that is, their ecology), he would need to consider not only the plants themselves but also the physical system that supports them. To make this easier, he coined the term "ecosystem" to describe the combined physical and biotic systems together. Tansley described these ecosystems as "the basic units of nature on the face of the Earth." Since then, scientists who study ecosystems have focused mainly on how energy and materials such as carbon, and nutrients that plants need such as nitrogen and phosphorus, flow and cycle through these systems. By the 1960s, aquatic ecologists had also begun to consider streams as ecosystems. They began to ask "Where do the energy and materials that support the plants, invertebrates, and fish in streams come from, and where do they go?"

We all know that plants need sunlight and carbon dioxide to grow and produce sugars (carbohydrates, which make up organic matter), and it is no different with plants in streams. But much of the plant life in smaller streams consists of the thin green slippery layer of algae that grows on rocks and gravel in shallow water. This thin film of algae, mixed together with bacteria and fungi, is called "periphyton" (*peri-* means around, and *phyton* means plant) or simply the "biofilm." However, the small streams that many ecologists first studied are deeply shaded by riparian trees and shrubs for most of the summer, so algae do not appear to be abundant during the plant growing season. As a result, ecologists began to wonder about the source of food for the stream invertebrates that feed fish, since most of these "bugs" eat plant material. Where could the carbohydrates and nutrients come from to fuel the ecosystem? It turns out that, like manna from heaven, some of the food falls from the sky. And, some of it actually flows in from beneath the ground.

Every fall, deciduous streamside trees and shrubs shed their leaves, which carpet the forest floor in beautiful yellows and reds. Some leaves swing and twirl down directly into the stream, but more blow in from the side when they become dry, during fall and early spring. A yearly budget of the organic matter that entered a small shaded New Hampshire stream showed that 86 percent of it came from dead leaves, wood, and other plant parts that fell or blew into the stream. Only a tiny fraction, much less than 1 percent, came from the growth of living plants (mainly aquatic mosses in that stream). The rest came in as "dissolved organic matter," such as the tea-stained leachings from dead leaves as rain percolated through the forest floor and into the groundwater that fed the stream. In streams shaded by conifers, the needles and wood also eventually fall and supply organic matter to streams.

But Charley and I learned from Ken Cummins that what drives the ecosystem changes as the stream moves from its headwaters to the middle and lower reaches. Trees and shrubs can't shade the entire width of larger streams and rivers, so more sunlight reaches the stream and grows more periphyton and other aquatic plants. As a result, these plants supply more of the food in the middle reaches, and dead leaves and wood supply less, in part because more of the leaves and small branches are washed away in these larger streams. All of this dead and living plant matter in the headwaters and middle reaches eventually becomes "processed," either chewed up by stream invertebrates or ground up by sand and gravel into tiny particles, and flows downstream. In larger rivers, the water also becomes muddy (or "turbid," in technical terms) so less light can penetrate to grow plants. As a result, much of the organic matter that feeds invertebrates in large rivers comes from the tiny particles that flow from upstream. By the late 1970s, when Charley and I were enrolled in Stream Ecology, these ideas—this theory about the continuum of changes in the ecosystem from small streams to larger rivers—was just being created and published by Ken Cummins and his colleagues. It became known as the River Continuum Concept (RCC) and has been a foundation for the canon in stream ecology ever since.

Like most theories in science, the RCC has been challenged by other scientists. For example, what happens in small headwater streams that flow through grasslands or above timberline where there is little shading from riparian trees or shrubs? Other scientists predicted that there would be more algae in streams of these open landscapes than in deeply shaded headwater streams, and that this would be a more important source of food for invertebrates than dead leaves. Indeed, this is the case, and was deduced from the original theory. Overall, the RCC has continued to be a resilient theory in many aspects, surviving several such "exceptions that prove the rule."

However, the main idea that most invertebrates in headwater streams eat dead leaves has required modification, because of new techniques that allow scientists to measure where the carbon in stream invertebrates and fish comes from. It turns out that there are different stable forms of the carbon atom, called "stable isotopes," and the mix of these is different in stream algae than in tree leaves. When scientists analyze these mixtures in stream invertebrates, they find that most are "built" mainly from the carbon in algae, except in very small headwater streams where deep shade really does slow the growth of these plants. However, a substantial amount of carbon in these shaded streams also comes from dissolved organic matter that enters in groundwater, is "eaten" by bacteria that make up part of the biofilm and is, in turn, scraped up and eaten by invertebrates.

In the end, even though dead leaves and wood are the most abundant sources of organic carbon in streams, invertebrates prefer to eat algae and bacteria that are more easily digested and yield more nutrients than dead leaves. They like the soft bread rolls (more succulent and easily digested algae and bacteria) better than the tough rye crackers (dead leaves). As a result, these insects keep the biofilm mown down to a thin layer. However, like your lawn during summer, it grows back very quickly and, ultimately, supplies more of the carbon and nutrients that end up in the insects than do the dead leaves. In contrast, the leaves and wood are processed and disappear slowly, so they appear to be much more abundant than the algae.

A Role for the Insects

Most fish in streams throughout the world do not eat leaves or scrape the biofilm from rocks, although a few do, primarily in tropical streams and rivers. So, most fish can't reach the energy and nutrients that plants store as carbohydrates in this way. The main link between the energy and nutrients in plants and bacteria and the food needed by fish is found in the stream invertebrates, which do eat both live biofilm and dead plant material like leaves, needles, and even wood. Like many stream ecologists, Ken Cummins was most interested in aquatic insects, the tiny larvae of mayflies, stoneflies, caddisflies, midges, and other insects that live in streams (see figure 6). These larvae eventually undergo an amazing transformation, in some cases similar to caterpillars that become butterflies. They leave their aquatic habitat and emerge from streams as flying adult insects, which then live for only a few days or weeks in the streamside forest or grassland. These are the clouds of diaphanous midges and mayflies that attracted my attention in early spring and summer as I first worked along Michigan streams.

Most streams and small rivers in the temperate latitudes support between five and thirty species of fishes, few enough to learn all of them by name. However, there are often fifty to a hundred species of aquatic invertebrates in thc same stream reach, so ecologists have found it easiest to put them into groups based on the roles they play. Some graze on the thin biofilm, like tiny bison or wildebeest, and are termed "grazers" or "scrapers." Others called "shredders" chew up the decomposing leaves to get not only the carbohydrates but also the microbes growing on and inside the leaves, which may also provide important nutrients. Still others, labeled "collectors," gather or filter out the tiny scraps created by all this mowing and chipping that flow downstream. Finally, a fourth group of invertebrate "predators," which may grow up to two inches long, eat insects in the other three groups. Most of these are bottom-dwelling, or "benthic" insects, which cling to the streambed or hide in the gravel rather than swim freely about.

Figure 6. Top: A mayfly nymph grazing on the biofilm growing on the surface of a rock on the streambed (genus *Epeorus*, a grazer). Bottom: A caddisfly larva that builds a web to filter food particles from the flowing stream (genus *Arctopsyche*, a collector). Images by Jeremy Monroe, Freshwaters Illustrated (used with permission).

Working from the theory that the main source of carbohydrates changes from dead leaves to biofilm as streams grow larger, stream ecologists reasoned that these invertebrate groups should also change along the riverscape. Shredders are usually most abundant in headwater streams where leaf input is high, whereas grazers are more abundant in mid-reaches where more biofilm grows. Downstream reaches are dominated by collectors, because most organic matter is in the form of tiny particles. Predators occur in low abundance throughout.

These aquatic insects have very interesting life cycles, which may last from only a few weeks to several years, but several common features can make them relatively easy prey for fish. For example, many of them drift downstream, either because they are swept away by the current or leave to avoid predators, or because they become too crowded for the food or space available. Trout, charr, and many other stream fishes focus on these drifting invertebrates as a food source. In addition, when adult aquatic insects emerge from the stream, they must leave the protection of gravel crevices where they have been hiding and either swim or float to the water surface or crawl out onto the bank.

Many insects drift and emerge at dusk or during the night when they are less visible to fish. One can imagine how natural selection would cause these behaviors to evolve, since few of those that emerge during the day would survive. However, many still fall prey to fish when the light is low. This is because many fish species can see very well even in low light. In addition, all fish have a sixth sensory system called the "lateral line" for detecting vibrations in the water, such as those created by invertebrates that are struggling to emerge from the stream surface. Stream invertebrates typically live only a few days as winged adults, and most then mate in swarms above the stream. After mating, the adult female insects lay eggs in the stream and die, and males may fall spent on the water surface, again making them easy prey for fish.

Surprisingly, although most biologists and anglers once thought that fish ate mainly these drifting and emerging aquatic insects, we now know that the terrestrial invertebrates that cross the boundary and fall into streams are equally important sources of food for fish. Although grasshoppers, or their imitations, are a favorite lure used by trout anglers in late summer, fish biologists who study what fish eat often ignored these and other terrestrial insects they found in fish diets, such as bees, wasps, ants, beetles, bugs (true bugs, like cicadas and leafhoppers), flies, and caterpillars of moths and butterflies. This was mainly because they found them in fish stomachs only on certain occasions, whereas the aquatic insect larvae that drift were always present. For example, during an entire summer of sampling fish to study their diets, consisting of perhaps a total of eight days on each stream studied (two days per month over four months), one might find large numbers of these "terrestrials" on only one day. They tend to come in waves, especially on warm windy summer afternoons when these insects are most active and more likely to be blown or fall into streams. Many biologists simply ignored these terrestrials because they were so sporadic, focusing instead on the more predictable aquatic insect prey.

However, we now know from studies throughout the world that these terrestrial insects generally make up about *half* the total weight of food eaten by fish in streams during summer. One reason is that terrestrial insects like grasshoppers and bees are about ten times the weight of the average aquatic insect larva, so a pulse of them on certain days creates a big feast for fish. Indeed, Shigeru Nakano and his colleague Masashi Murakami showed that the terrestrials that fell into a Japanese stream, mainly during summer and early fall, made up nearly half of the total energy needed by trout and charr throughout the entire year. It was the movements of emerging adult insects into the riparian zone, and these terrestrial insects that fell from the sky, that led Nakano to his groundbreaking new idea about how streams and forests are linked together, and why they are critically important to each other.

Crossing Boundaries

Shigeru Nakano and his colleagues made some of the most surprising discoveries in all of stream ecology in the studies they conducted on one small stream in southwestern Hokkaido during 1995 to 2001. They found that cutting off these nearly invisible connections made by tiny invertebrates crossing the aquatic-terrestrial boundary between streams and forests or grasslands can cause half or more of the animals to disappear, and completely change the way these ecosystems look and operate.

When Nakano and his students simply stretched clear plastic or nylon mosquito netting over agricultural greenhouse frames to cover reaches of Horonai Stream (see inset in figure 2 for location) they found that half the trout and charr left. In two other "greenhouse studies" on the same stream, female bats that rely on emerging insects for protein during pregnancy in early spring virtually stopped feeding in the streamside forest, and seven of ten riparian spiders left. In a fourth study, when Nakano and his colleagues used mesh fences to enclose Dolly Varden in reaches covered by greenhouses to prevent the charr from leaving, the fish switched to feeding on the tiny aquatic insect larvae from the streambed, just as we had found in Poroshiri Stream when drifting insects were filtered out. In turn, this caused the thin layer of biofilm to blossom into large clumps of algae and turned the streambed a dark green. When the terrestrial invertebrates were cut off, the fish decimated the aquatic grazers and the plants grew unchecked.

Is this some sleight of hand, some trick whereby bats and fish develop phobias toward transparent greenhouses? Or do we humans simply not see the important linkages between streams and riparian forests that, when cut off, can cause such drastic effects? Ecosystems are often driven by small forces that accumulate over time, like millions of earthworms slowly but inexorably processing dead plant matter into the soil on which humans depend. Likewise, the daily exodus of emerging insects struggling off the surface of Horonai Stream that Nakano studied amassed more than a half pound of aquatic insects in each 100-yard section of riparian forest during the summer season. In return, the riparian forest supplied nearly five pounds of terrestrial insects to the 12-foot-wide stream along the same reach. It's no wonder that so many fish, birds, bats, and spiders line up to feed on this "subsidy," which is an important windfall (or landfall) for them.

Figure 7. A mesh greenhouse used to cut off movements of adult aquatic insects from Horonai Stream to the riparian forest, and terrestrial forest invertebrates into the stream. Image by Colden Baxter (used with permission).

Why hadn't scientists told us about these connections before? They seem critically important, and have big effects on fish in streams and animals like birds in riparian forests. Didn't they learn in school that basic lesson of all ecology, that everything is connected to everything else? In truth, it took a long time for scientists to discover these connections for one simple reason—because we often isolated ourselves in rather narrow disciplines. The old adage is that experts (especially academic scientists) work entire careers to learn more and more about less and less until eventually they know everything about nothing. Many field ecologists, for example, have a passion for certain animals such as birds, or fish, or insects in streams. They train for years to learn to identify their many forms, and how to study them, and then work only with this one group. The need for scientists to also become experts in statistics and mathematical models also limits their ability to master the natural history of many diverse groups of animals, especially those that cross boundaries to other ecosystems. Biologists who work with amphibians are one exception, and some make a career of following their subjects from stream or pond to forest and back.

But Shigeru Nakano was different, perhaps because as a Japanese ecologist he viewed the stream and riparian ecosystem more holistically. By 1994, he had become well known in Japan for his research on charr, and in 1996 he landed the position of Director of the Tomakomai Experimental Forest of Hokkaido University, a forest preserve that included the Horonai Stream watershed. Very early each morning he drove to work along the forest road near the stream, and would sometimes stop to make observations. He and his collaborator Masashi Murakami noticed that in springtime, warblers, wrens, and other forest birds were feeding actively along the stream banks, and even in the spaces underneath them, for tiny emerging mayflies, caddisflies, and stoneflies. Nakano's ideas were also catalyzed by a scientific paper he had read, by an American ecologist an ocean away who had studied birds foraging on emerging aquatic insects along a stream in the Kansas prairie. It is not uncommon for one's research to spark a flame in someone else's imagination in a distant land, perhaps some years in the future, far beyond what you could have imagined when you wrote the original paper.

Few scientists I know could match Nakano's drive to undertake so many large-scale studies and collect the data needed to convincingly answer scientific questions. For example, in one study he and Murakami and a few colleagues made over 13,000 careful observations of birds feeding on insects in the riparian zone during a fourteen-month study. They worked every day (except when it rained) to follow individual birds with binoculars and record when they captured aquatic versus terrestrial insects. Based on this Herculean effort they discovered that, on average, fully a quarter of the annual energy needs (calories) for the ten species of birds that feed in the riparian forest were supplied by the small adult aquatic insects that emerged from Horonai Stream. Likewise, sampling fish every month and flushing the stomach contents of over 1,400 of them showed that nearly half of the annual energy for the five species of fish present was supplied by terrestrial insects that fell into the stream. When it was published, this study, along with a previous one using greenhouses, attracted the attention of many scientists worldwide and elevated Nakano to the status of an ecological rock star. As the first paper was being published, he earned a job offer from Kyoto University, the equivalent in Japan of Harvard or Yale, and chances to visit and collaborate with the most famous ecologists worldwide.

But where did Nakano get the idea to cover the stream with a greenhouse to cut off these movements of insects from stream to forest, and from forest to stream? During the year before he moved to Tomakomai, Nakano began to ponder his new position and opportunities for research funding to study how to conserve biodiversity in Japan. Most stream ecologists had studied why forests and grasslands were important to streams by measuring the leaves and wood that fell into them and, in a few cases, terrestrial insects. But Nakano asked the converse question—"Why are streams important to forests, and the many different species we find there?" To test this idea convincingly, he realized he needed to cut off these insects from reaching their destination. When I wrote to his graduate students and colleagues from the mid-1990s, they replied that Nakano was strongly influenced by the field experiment I had designed to cut off drifting insects in Poroshiri Stream using nets. Then, driving across the rural Hokkaido landscape he noticed the many greenhouses on small farms there. As my colleague Colden Baxter later quipped, Nakano probably thought, "My stream would fit right inside one of those!" This simple but powerful idea led to four main studies with greenhouses by Nakano and his colleagues in Japan by 2002, and many more by scientists in other countries during the decade after his work was published. Nearly all have shown that cutting off these movements by insects has the same drastic effects.

Nakano was not the only scientist to have these ideas about the importance of insects moving between streams and their riparian zones. Great minds think alike. My colleague Mary Power at the University of California at Berkeley had also been studying how bats foraged on insects emerging from streams, and measuring how they deposit the nutrients from these prey as guano into upland forests far above the Eel River in northern California. John Sabo, a PhD student working with Mary, was also inspired by her ideas about river-watershed linkages. Sabo set up plastic shields along the river's edge in 1997 and 1998 to exclude emerging insects, and measured the abundance and growth of western fence lizards that live on the dry cobble bars along the river during summer. Similar to Nakano's results, Sabo and Power found that four of ten lizards left the riparian zone when insect emergence was cut in half (although not eliminated). Not only that, lizards enclosed in plots that received insect emergence grew seven times faster than those enclosed in plots where insect emergence was reduced with the shields. But surprisingly, the two groups of scientists knew nothing of each other's research. Power and her colleagues were also studying spiders, like Nakano, and the two groups were converging on the same findings. Little did they know that within a few years their paths would cross.

What about Fish?

So far we have been considering "food webs" that link streams and their riparian zones. Ecologists use this term simply to describe "who eats whom," which also determines the important pathways along which energy and nutrients flow in ecosystems (see figure 8). We know that both aquatic and terrestrial invertebrates are important because they not only provide food for predators within the systems in which they first live, but also cross the terrestrial-aquatic boundary to feed predators in the other habitat. But, what about the fish themselves? Do they have effects other than simply eating these invertebrates? Are they important in stream, river, and riparian ecosystems?

From the perspective of the ecosystem, fish make up only a tiny fraction of all the organic matter, and energy, even if we consider only the portion within the stream. For example, in Ken Cummins's Stream Ecology course, Charley Dewberry and I learned that all the "macro-consumers"

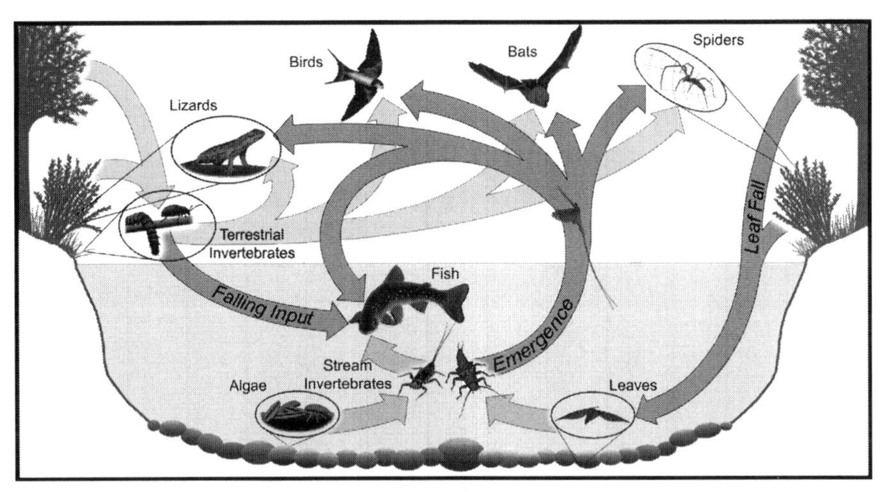

Figure 8. Diagram of a food web in a linked stream-riparian ecosystem. Arrows show the feeding relationships among animals and plants. For example, stream fish eat both aquatic insects and terrestrial invertebrates that fall into streams. A substantial fraction of the diet of birds, bats, lizards, and spiders in the riparian forest can be adult aquatic insects that emerge from the stream. Image by Jeremy Monroe, Freshwaters Illustrated (reprinted from Baxter et al. 2005, with permission).

(all the stream invertebrates and fish) in the small New Hampshire stream described above required only a miniscule amount of the energy flowing through this ecosystem to support their populations—far less than 1 percent. In fact, the scientists who developed the River Continuum Concept considered fish so insignificant that the small icons they used for the fish in their diagram were only about the same size as the icons for shredders, collectors, and predator invertebrates. After all, how important can organisms like fish be to a food web when they make up such a small part of the whole ecosystem? But as it turns out, fish have big effects, which cascade from the top down to the very bottom of food webs in streams.

The South Fork of the Eel River has been one of the natural laboratories where stream ecologists discovered the role that fish play in streams, even before Nakano began working on this topic in Horonai Stream in Hokkaido. The South Fork dissects the Coast Range of northern California, flowing cool and green through a deep canyon bordered by old-growth Douglas-fir and coastal redwood forest. Long gentle riffles glide into deep pools during summer, but winter rains can bring huge floods that winnow silt and scour periphyton from the gravel and cobble in the riverbed. Steelhead, an oceangoing form of rainbow trout, ascend the river to spawn as these winter floods peak, and their young live there for two to three years before returning to the ocean to feed and grow to adulthood. Several other minnow-sized species of freshwater fish also inhabit the river. It is the effects of these fish on stream ecosystems that Mary Power, one of the most renowned stream ecologists, and her students have studied.

I met Mary at a scientific meeting in the 1980s, the blur of years when my children were very young and I was busily working to gain tenure at my university. Although my background had been more "applied," in fisheries biology, and hers more "basic," in aquatic ecology, we became lifelong friends after our first meeting. From the very start of my career, I had been keen to cross the boundary from fisheries toward ecology. In the 1980s, this was a somewhat uncomfortable "no-man's land," where colleagues in either field might disown you for sailing beyond the horizon of the topics considered important to study. Fortunately, Mary was a welcoming member of the community of ecologists studying streams, and offered to show me the ropes and introduce me to the right people. She has done that for me my entire career.

Most of the early work on the role of fish had been done in lakes. Aquatic ecologists at the University of Wisconsin, the powerhouse for lake ecology in North America, had studied whether adding or removing fish could make a difference to the invertebrates and plants in lakes. In these relatively stable ecosystems (compared with flowing streams), small fish eat tiny crustaceans called "zooplankton." These tiny animals that live in the open waters of lakes are the aquatic equivalent of cows, and graze on even smaller "phytoplankton," the tiny plants suspended in the surface waters. So, when lakes have many small fish, they filter out and eat the zooplankton, allowing the phytoplankton to "bloom." In an extreme case, the lake can turn as green as pea soup. However, when larger predatory fish eat the small ones, or in lakes without fish, the zooplankton populations explode, happily eat the phytoplankton, and the lake becomes clear.

These effects also depend on the amount of nutrients present in the lakes to fertilize the phytoplankton. Indeed, these nutrients (primarily nitrogen and phosphorus) are the same ones we spread on our lawns to make them green. Even so, these cascading effects from fish to microscopic plants that completely change the way lakes look are predictable enough that ecologists have named them "trophic cascades." The scientific term "trophic" refers to the feeding relationships in the food web. We now know that not only fish, but also other predators like wolves and sea otters, can cause trophic cascades and turn the world green. Wolves prey on herbivores like elk and deer in mountain valleys, and sea otters eat sea urchins on ocean coasts. These herbivores, in turn, eat the plants in these ecosystems. Put simply, more predators mean fewer herbivores, and so more plants.

But no one really believed that the same things could happen in streams and rivers, especially those, like the Eel River, that flood every winter. Floods and droughts were thought to have much stronger effects on stream algae and invertebrates than fish ever could. Mary Power wondered whether the young steelhead and other small fish in the Eel River could eat enough invertebrates to affect periphyton on the rocks via a trophic cascade, working from the top of the food web down. She planned an ingenious field experiment to find out, which showed that fish had stronger effects than most stream ecologists had imagined.

One way to determine the effects of fish in streams is to build cages to either exclude the fish from a small part of the river or to enclose them there. When Mary conducted this experiment in summer 1989, she found that the fish had drastic effects, but in the opposite direction than one might expect based on the simple three-level food chain described above for small lakes. Simply adding 2-inch-long fish (steelhead fry and the main species of minnow) to her cages caused them to be nearly barren of algae, whereas in those without these fish the algae bloomed in great green clumps by late summer. How could this be? Is there some reason the theory appeared to fail? Fortunately, her experience and expertise with invertebrates and algae helped her discover the answer.

The reason for these unexpected results was that many predatory invertebrates and small fry of other fishes in the river (all about 1 inch long by the end of the experiment) formed another link in the food chain, causing the trophic cascade to increase to four levels. So, the young steelhead and adult minnows (level 4) ate large predatory invertebrates and small fish fry of other species (level 3), which released the smaller invertebrate herbivores from predation (level 2; midge larvae in this case), and these ate the algae (level 1). Then Mary's results fit perfectly with the trophic cascade theory, which predicts that in food chains with an even number of these "trophic levels" (in this case, four levels) plants like algae should be depleted. Likewise, when she reduced it to only three levels by excluding the steelhead and adult minnows, the predatory invertebrates and fish fry decimated the herbivores and the algae bloomed, just as the trophic cascade theory predicted should happen in chains with an odd number of levels (see figure 9).

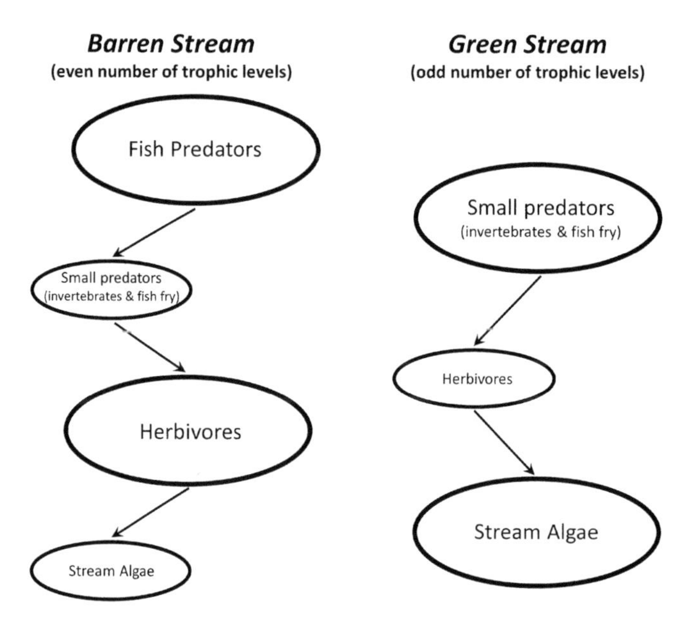

Figure 9. Effects of fish predators in river food webs, based on the theory of trophic cascades. Larger ovals indicate greater abundance of organisms, and smaller ones less. When large fish are present, creating a 4-level food web (an even number of trophic levels), they eat small predators, releasing herbivorous benthic invertebrates which, in turn, graze stream algae to low levels and create a barren stream bed. In contrast, when only small predators are present, such as invertebrates and fish fry, these eat the herbivores and reduce their numbers. In turn, this allows algae to bloom and turns the stream green (after Power 1990).

Nakano and I, and every other stream ecologist, had read about this famous experiment by Mary Power, which was published in 1990. It has long been a classic study that we present to every student in our courses on aquatic ecology and fish biology. But Nakano wondered how the terrestrial insects that fall into streams might alter this story. He knew from our previous work that Dolly Varden charr are adept at eating the invertebrates that graze on periphyton in Hokkaido streams, but he also knew that the terrestrials made up the other half of their diet. Based on this food web, Nakano began to wonder whether these terrestrial insects might play a role in controlling how much algae grows on the surfaces of rocks in the stream, by changing fish foraging. Although at first it seemed far-fetched to think that tiny insects that fall from the forest into the stream could have such strong effects, Nakano imagined an experiment that he could use to find out. He decided to cut off the terrestrial insects from reaching the stream using greenhouses. He set out to test whether, indeed, "everything is connected to everything else."

Every one of my colleagues remembers the day when they first saw Nakano's paper in 1999 from this work because, for most, it changed their world. LeRoy Poff, who teaches Stream Ecology at my university, read it only a half hour before his lecture on stream food webs, and immediately changed his lecture because the results were so astounding. John Sabo, having just the year before conducted his own PhD research with shields that reduced the insects emerging from the Eel River into the riparian zone, was very surprised that someone else had used a similar technique, but was also very impressed with the research. When Nakano installed greenhouses to prevent terrestrial insects from falling into sections of Horonai Stream, the Dolly Varden immediately switched to feeding directly on stream invertebrates that graze periphyton and, by the end, cut the total weight of these bugs by half. With half the grazers gone, the periphyton bloomed, increasing by 50 percent (Horonai Stream had a three-level food chain, unlike the Eel River). Before this, no one had realized the power of these small terrestrial insects that rain into streams to alter how these ecosystems look and operate, especially through indirect effects, caused by fish, that cascade through these food webs. After all, terrestrial insects don't graze on stream algae, but if they disappear the result is that the algae turn the stream bottom an entirely different color and texture.

The connections that Shigeru Nakano and Mary Power and their colleagues discovered through the invertebrates that move between streams and forests are just some of the many connections that we now know are important to these ecosystems. For example, tiny rivulets that are too small to support fish nevertheless carry invertebrates and dead leaves into larger streams, which are important sources of food and nutrients for the fish and invertebrates downstream. In the opposite direction, fish migrating into streams and rivers from lakes and oceans also carry nutrients upstream that enrich these flowing waters and their riparian forests. For example, salmon migrating up rivers from the Pacific Ocean carry large amounts of "marine-derived" nutrients, principally nitrogen and phosphorus, into streams. Many of these salmon are dragged into riparian zones and eaten by bears, or scavenged by eagles and smaller mammals, and dead carcasses of those not eaten can become stranded in the riparian forest by floods. Eventually, the nitrogen brought by the salmon ends up in the soil, where it fertilizes riparian trees and shrubs. When runs of salmon, and the bears that eat them, are lost—such as when dams are built that block salmon migrations—then riparian forests are missing important nutrients that sustain their growth and productivity.

Even fish that do not migrate to an ocean can make important movements along streams, connecting one part of a watershed to another. But like the movements of tiny insects to and from streams, these peregrinations are often hidden, even to scientists. At about the same time Nakano and I began our studies of charr in Japan, I also stumbled on the importance of these connections while working with graduate students on streams high in the mountains of Colorado. Here, too, serendipity allowed us to make discoveries that changed the way we think about fish and streams.

Landscapes to Riverscapes

Wonder and excitement overtake me the first time I approach a stream I plan to study. As I hike down from the uplands, the phalanx of willows or alders guarding the riparian zone hold their arms high and pull at my arms and clothes as I clamber through their midst. Finally reaching the stream, I seek a vantage point high on a bank by a pool, eager for a broader view. Peering upstream and down, I take in the entire scene to the next bend in each direction. I strain to see the shapes of fish in the blue-green depths of the pool, their forms hidden by bubbles carried in the currents and distorted as rays of light are bent by the water. I gauge whether the stream and its channel are in good condition, or have suffered from human uses of the land and water. These few glimpses form my first impression of any stream. But then, to see more of the habitat, I must either clamber through the riparian zone again twice to survey the next short reach, or hike or snorkel some distance along the stream itself to explore the mysteries beyond the next bend.

This experience defines the difficulty of understanding streams. It is a rather different problem than the complexities of studying lakes. When approaching a lake, you can often see the broad expanse that you plan to study. The challenge there is finding the best way to capture and measure the abundance of the different organisms throughout this huge body of water. But from any single vantage point along a stream or river you can gain only a glimpse of the habitat, from one bend to another, which makes it difficult to even envision your object of study. In addition, the fish and invertebrates and plants you find there live in an ever-changing and continuously renewing flow of water, which comes from myriad places unseen, and travels to the sea beyond. This onrushing flow washes away any materials and organisms incapable of withstanding its relentless pull. But it also offers a hydraulic highway for any animal that can swim, drift, or even crawl, and allows it to move far beyond the short reach within view. How can ecologists hope to understand such a system that is constantly changing, and connected to the broad branching hierarchy of channels beyond?

With regard to fish living in stream hierarchies, a basic question asked by many early biologists is "How long is the reach in which fish live most of their daily lives?" This reach is known as their "home range." This basic knowledge is needed by fisheries biologists so that they can plan effective management. For example, it might seem ineffective to set restrictive regulations on fishing in a half-mile section of stream, or restore the habitat there, if the fish regularly move many miles along the stream course in their normal activities.

To measure home ranges, biologists typically captured fish from a series of five to ten consecutive reaches, often about 50 yards long each, and marked the fish using different marks for each reach. When they sampled the reaches again later, they usually found that most of the marked fish they recaptured had remained in their "home" section from one occasion to the next. This led them to believe that most fish lived within home ranges of only 20 to 50 yards their whole lives. However, the majority of marked fish often went missing and were never recaptured. When the biologists went looking for these fish by sampling other 50-yard reaches some distance beyond their entire study segment, they usually found no marked fish, or only one. So they concluded that, indeed, most fish must be relatively sedentary and stayed home. The missing ones probably had died or, because sampling is difficult, were simply not captured again but were presumed to be in or near their home reach. However, the first large-scale study my graduate students and I conducted after I arrived at Colorado State University helped change this view forever.

In the early 1980s, fisheries biologists in the U.S. Forest Service and Colorado Division of Wildlife had installed logs to form pools in more than a dozen small streams in the mountains of northern Colorado in hopes of increasing the number of trout for anglers. They reasoned from the previous research that trout are sedentary and have small home ranges, but need deep pools to survive, especially during the long winters when liquid water is scarce in these cold, high-elevation streams. Many shallow pools freeze nearly to the bottom, leaving little living space for fish. Shelter from logs and undercut banks is also important during late spring. Melting snow produces raging floods of icy water that can stress trout, which are often thin and wan after the long winter. To create more pools, the biologists placed logs straight across the channel and dug them into the banks. These formed a low small dam that created a pool upstream and caused the stream to scour out a deep plunge pool downstream and underneath the log (see figure 10). However, no one knew whether the logs actually *did* increase the number of trout, and so were worth the money to install. In addition, few biologists stopped to think about why the streams lacked pools in the first place.

I jumped at the chance to study this problem when it was suggested, and after two years of writing and rewriting the proposal I finally landed the funding. It was my first big break to develop a comprehensive study of trout in Colorado. The study design was simple in principle, but difficult to carry out in practice. We would find six different small mountain trout streams that had 500-yard segments with few natural logs and few pools, where the habitat consisted of mainly shallow "riffles" and "runs." Such streams were not uncommon in the mountains of northern Colorado.

The experiment consisted of dividing the 500-yard segment of each stream in half and installing ten logs in one of the two 250-yard study reaches. We flipped a coin to choose the reach for the logs in each stream at random, to make sure that we didn't skew the results by choosing the one where we thought trout might respond more to the pools we created. We measured the number of trout in each study reach for two years before the logs were installed, and then for six years afterward. This was a huge effort, which took the help of dozens of graduate and undergraduate

Figure 10. A typical log structure installed in a Colorado mountain stream to increase pool habitat for trout. High flows dig a plunge pool downstream and underneath the log, and a pool also forms upstream, both of which are important to trout during low flows and freezing temperatures in winter and high flows during spring snowmelt runoff (image by Kurt Fausch).

students. We expected the number of trout to increase in the sections with logs, but to remain about constant through time in those without logs. I chose 250-yard study reaches because they were five times the length of the home range of an individual fish, based on what earlier scientists had reported about trout movement. However, this assumption proved to be far too simple to describe how trout actually used the habitat in these streams.

Overall, the logs worked perfectly to create pools. And the number of adult trout in the sections with logs increased by about half, such as from 100 to 150 trout, which is a very large increase. However, the trout did not increase for the reasons we originally thought. Instead of increasing the winter survival of the trout living in the study reaches, as biologists had believed, the logs created new habitat that was quickly colonized by adult trout that were moving throughout the watershed, from short and long distances. Analysis of our first four years of data suggested that many more trout were moving than we expected, so PhD students Chas Gowan and Steve Riley placed fish traps at both ends and the middle of the 500-yard study segment in two of our six streams to measure how many fish were moving. They found that more fish moved through the traps in both directions during the hundred-day summer season than were found living in the sections when we sampled them in early fall. Our assumption based on previous studies that most trout would "stay home" was clearly wrong for these streams.

Chas Gowan conducted more detailed studies and analyses of the moving trout, which ultimately revealed why the logs increased trout abundance and why other biologists had concluded that most trout are sedentary. About half the trout marked do remain concentrated within a relatively short reach for some years, and so are available to recapture in a study like ours. But the other half can move long distances, often half a mile or more in these small streams that averaged about 12 feet wide, and disperse widely throughout the watershed in search of favorable places to live. They cross boundaries that take them far beyond the study reaches laid out and sampled by fish biologists, and so are never seen again. Moreover, because they disperse widely, few of these trout colonize any particular distant reach. Therefore, biologists who look for the missing fish in a few short outlying reaches recapture few or none, and are deluded into thinking that few moved.

However, the trout that range over long distances are also the fish that find new habitat, like the pools we created with logs, and quickly colonize them, just as we found. When one trout dies, or is caught by anglers, these immigrants quickly take their place. We recently returned to five of these six streams, twenty-one years after the logs were installed, to measure how the pools and the trout had fared. Surprisingly, every log was still in place and had formed a pool, and the number of adult trout was still just over 50 percent greater than in the paired reaches without logs. This study stands as the most comprehensive and among the longest ever conducted on the effect of habitat restoration for trout in streams.

But why do so many small streams in the Rocky Mountains have so few logs that create pools naturally? The answer is quite simple. All the trees within a few hundred yards of nearly all these streams were cut between 80 and 140 years ago to supply railroad ties and mining timbers. This was the period when the transcontinental railroad was being built across southern Wyoming and mining towns boomed in Colorado. These logs were sluiced down the streams and rivers themselves in the spring "runoff" flood, but only after all the natural logs and boulders that made pools were cleared from the channels. Sometimes a large "splash dam" was also built of wood, the logs were loaded into the pond when the snows melted, and the whole thing was dynamited to create an even larger flood to transport the logs. Even after 100 years or more, after the forests have grown back and look pristine, the channels are still artificially "clean." No trees have yet died and fallen in to create new pools, and the boulders lie idle aside the stream in the riparian zone. For most streams, the next fifty years will be critical, when the current forest will begin to supply new fallen trees to create pools. Perhaps then, trout moving in these streams will encounter pools more frequently than when we studied them. If so, then these restored habitats will support more trout for anglers and natural predators to seek.

This eight-year study had a profound effect on all our research afterward, and joined with other research to have an important influence on the entire field. Gowan and Riley and I published a paper with Mike Young, another colleague studying trout movement, pointing out the flaw in arguments by other scientists that most trout and other stream fish are sedentary. In fact, other biologists had often found that fish quickly recolonized stream sections that previously had gone dry, one of several observations that contradicted the original theory that most stream fish were sedentary. New studies by us and others focused on such fish movements, adding further evidence that they are important. Later, I found kindred spirits in Christian Torgersen and Colden Baxter, both PhD students at Oregon State University working with professor Hiram Li. Together we synthesized a new general theory explaining how and why fish in streams use habitat at much larger scales than originally believed, and we coined the term "riverscapes" to describe the concept. Many stream fish ecologists now design their studies based on these ideas.

What if we could zoom out to view an entire watershed, as with Google Earth, and see where all the fish are going throughout their lives? We would discover that across different seasons fish move to use several different complementary habitats, all dispersed throughout the riverscape. For example, adult fish often move to specific locations to find the right type of gravel or plants in which to lay their eggs. When the eggs hatch and become larvae, the tiny fish often drift some distance downstream and colonize shallow productive backwaters where they feed and grow.

When they grow larger, these juveniles can swim against stronger currents and move into the main channel to feed. However, juvenile and adult fish often must move to distant refuge habitats during periods of summer drying or winter freezing, and then back to productive feeding habitats after this "crunch" period passes. These complementary habitats may be rather close together, within 10 yards of each other, or may be many miles apart. We now know that even some small fish only 3 inches long may need to traverse 50 miles or more of river, crossing many boundaries to find all the habitats they need to complete their life cycle. This has made understanding what habitats different stream fish need, and how to conserve them, much more challenging, but our answers are also much more realistic and useful to those who manage these fish and the watersheds in which they carry out their lives.

Ecological Surprises

In the end, our studies of how fish move to use habitat over long distances in riverscapes, and the research by Shigeru Nakano and Mary Power and others on the connections between streams and their riparian forests through tiny invertebrates, have revealed hidden connections that are critical to supporting the animals that live in these linked ecosystems. Without the terrestrial insects that fall into streams from riparian shrubs and trees, fish are missing half their diet and many will leave for greener pastures, or have stunted growth or lower survival if they stay. Without the adult aquatic insects emerging from streams into forests and grasslands, birds, bats, lizards, and spiders will either starve or leave. Without free access to deep, stable pools with abundant logs or undercut banks, and clean gravel riffles in which to lay their eggs, trout will either die during winter or be unable to spawn, and their numbers will decline. Anything we do as humans to disrupt these movements of invertebrates and fish across these boundaries will prove to reduce these animals that we may care about, as surely as we humans would also be reduced if our food and shelter were denied us. And this idea about how human actions can disrupt these movements and affect both streams and their riparian forests would prove to be the key to our ultimate triumph over tragedy in Baja California.

Chapter 4
Tragedy in the Sea of Cortez

"Hello, Kurt? This is Yoshi. I'm in Tijuana, Mexico." The sound of Yoshi Taniguchi's anxious voice on the phone took me by surprise, because we had always communicated by e-mail. "Nakano has been lost in a boat accident in the Sea of Cortez with Gary Polis. Professors Abe and Higashi have drowned, along with Polis and his lab manager, but Nakano has not been found." Suddenly, my heart was beating in my throat. "Can you come to Baja?" Yoshi asked.

I knew that Shigeru Nakano had been planning to travel to California with Professor Masahiko Higashi to visit with Gary Polis, a professor at the University of California at Davis, and Mary Power at UC Berkeley. Two days before he left, Nakano had sent me comments on the manuscript we were writing together and apologized that they were hurried because he was preparing to leave. He had recently moved to his new position at the Center for Ecological Research (CER) at Kyoto University, and the trip was planned to help foster their new Research Institute for Humanity and Nature. For his part, Nakano wanted to see the field sites of these two famous American ecologists, because each of them had been studying how aquatic organisms and ecosystems are linked to terrestrial ones across the land-water boundary. And, Nakano and I had recently discussed how humans might affect these connections.

As an example of these linkages, Gary Polis and his collaborators had made the surprising discovery that plants and animals on small desert islands just off Baja California in the Sea of Cortez are strongly affected by the dead kelp and carrion of marine animals (from squid to whales) that wash up on their beaches. When they collected this sea wrack and dried it, it averaged more than sixty pounds for each yard of shoreline each year. All this dead material supports a great abundance of different kinds of small invertebrate scavengers, which are then eaten by spiders and scorpions that live near the shore. In turn, these predatory spiders also eat other insects that live on, and chew up, island plants. So, similar to Nakano's findings about connections between streams and their riparian forests, Polis's group had discovered that connections between the sea and the shore could strongly affect the predators and plants living on islands by these indirect pathways. In a similar study on the coastline of Namibia, Africa, Polis and his colleagues found that the higher spider abundance near the shore reduced plant-chewing insects, so the plants living there had fewer damaged leaves and fared better than those inland because of this unique type of trophic cascade.

Shigeru Nakano wanted to see the desert islands off Baja for himself, and Polis wanted to show him the organisms and habitats of this starkly beautiful place he had been studying. Gary had heard about this young scientist from Japan, who just the year before had published a breakthrough paper on connections between streams and forests. Together with several scientists and technicians from Polis's laboratory and a large group of other students and Earthwatch volunteers (a nonprofit environmental group), Nakano and his two Japanese colleagues drove south to Bahia de Los Angeles, a large bay about halfway down the Baja Peninsula where Polis had been working for many years. They planned to spend a week conducting research with Polis's group. On a Monday morning, March 27, 2000, seventeen scientists and volunteers set out in two boats for Isla la Ventana (Window Island, named for a rock archway), about four miles from the marina and two miles from the nearest shoreline.

After Yoshi's phone call, I was shocked, and worried. I had to tell him that I was not able to come to Baja to help, because I was in the middle of teaching a new graduate course and preparing lectures nearly every waking hour. He had traveled to Mexico with Nakano's wife and parents and another senior faculty member from CER. They all waited anxiously while the search progressed. But what could I do to help, from such a long distance? I called the office of Dianne Feinstein, the senior U.S. senator from California, to ask what could be done, and was referred to the Coast Guard. The Coast Guard official said they had been taking part in the search, but informed me that they search only for survivors. After all hope of human survival is gone, using calculations based on water temperature and other conditions, they cease searching and do not attempt to recover bodies of victims.

I felt helpless. I checked online and found that reports about the accident had started appearing in the California media. On Tuesday morning, the day afterward, four survivors had been found on nearby islands, and the bodies of three victims were recovered. The Coast Guard was reportedly optimistic at first about finding other survivors because the waters were relatively warm (65°F), and there were many islands and many tourist boats that might locate survivors. Then Polis's body was found Thursday morning, hidden in a rock crevice on an island, after being spotted by a private helicopter pilot. But by Friday, Nakano had not been found. Despite my worry, I was confident that he was alive somewhere on one of the many small and large islands that dot the bay. Nakano was strong, and I knew from our work together in the remote mountains of Hokkaido that he was more tolerant of cold water than I. Besides, he knew how to take care of himself under difficult field conditions. Surely, I thought, he swam to an island, could survive a few days, and would soon be found.

Yoshi called again with more information, and I was able to piece together other details from news on the press conference with the survivors from UC Davis, held in Sacramento on Thursday night when they returned. About noon on that Monday, a storm had blown up suddenly while the group was on the island, and they had decided to return to the mainland. The wind blew in gusts so strong that it was difficult to stand, one survivor said. After motoring about 500 yards, the two boats lost sight of each other as the sea began to swell to 6 feet, and then to more than 10 feet. The one piloted by a local guide returned safely to shore. However, large waves came over the front and sides of the other boat carrying Polis and Nakano and seven others. They attempted to pull the drain plug and keep moving to drain the boat, but a large wave swamped the motor and it stalled. Nakano used an oar to attempt to flag down the other boat. Within a few minutes, another wave capsized the boat and the heavy seas then "pounded" the nine crew members relentlessly, according to Gary Huxel, a postdoctoral researcher in Polis's lab and one of the survivors.

Heroism is a quality that others assign after the fact, but several of the people on the boat who died that day thought more about others than about their own survival. Michael Rose, the lab manager, and Gary Polis worked hard to keep the crew together near the boat, but the slippery hull offered no handholds and each large wave washed them off into the open sea. After a few hours, Polis apparently had a heart attack in the water and died. Shigeru Nakano put his own life jacket on Takuya Abe, his eldest colleague, and helped Huxel tie a line around the boat for others to hold on. Survivors reported that as the sea raged and washed the group apart, Shigeru swam out repeatedly to bring people back to the boat and keep them together. Gary Huxel credited Nakano with saving his life.

Such violent westerly squalls, called *chubascos* by the locals, can blow up unexpectedly and wash boats off their moorings. After about three hours of battling the ocean through the storm, Huxel and three other graduate students and volunteers decided to swim for an island, fearing hypothermia. Nakano apparently decided to stay with his elder Japanese colleagues, who probably could not swim. He was also likely exhausted from helping other members of the crew. At the last, Huxel saw him floating near the boat with a seat cushion, as the boat and remaining crew members were being pulled by a strong tidal current out from the bay and toward the main Sea of Cortez. After several more hours of swimming, the four survivors reached two different islands, and spent the night there exhausted and sick from swallowing seawater. Mexican authorities discovered them Tuesday morning when they began searching at first light and recovered the boat and three bodies that day. The crew of the other boat reported the accident Monday evening. They returned to search for their colleagues, and asked other boaters, but found no sign of them. Nevertheless, they were confident that Polis must have pulled into a cove to wait out the storm. He had worked in the Bahia for many years and knew the sea, they thought.

Days passed, and Nakano still had not been found. The islands have few trees, so after all of them were searched and more time passed, there seemed little hope that he was alive. The Mexican Navy and U.S. Coast Guard began the search with two helicopters, five ships, and two hundred soldiers, and covered a 40-by-40-mile area of ocean. But on Thursday night, after more than three days, the Coast Guard halted their search, considering it unlikely that he was alive. The Mexican Navy continued to search, and soldiers combed miles of beaches for three weeks. Nakano's body was never recovered.

About ten days after the accident I received a call from UC Davis asking whether I could speak at a memorial service on behalf of the Japanese scientists. They had learned that I was a close colleague of Shigeru Nakano's. I agreed, even though I knew that Yoshi and Nakano's family felt that the university had not done enough to help them in their search for Shigeru in Mexico. Although they were able to speak to the survivors in San Diego as they traveled south, no one from UC Davis had gone with them to help. At the memorial, Gary Huxel also spoke, and was still badly shaken. His voice quavered as he told of Nakano's heroism, and again credited him with saving the lives of the survivors. I told the audience of about five hundred people that Nakano was a superb diver and field biologist, the best I had ever known. But I knew from personal experience during grueling, difficult field research in the mountains of Japan and Montana that he would never have left his friends to save himself. His courage and actions are embodied in the passage from the Bible about there being no greater love than to lay down one's life for one's friends (John 15:13). Drawing from their deep wisdom and grace, his parents Hiroshi and Hiroe Nakano later said that his sacrifice was "the right thing to do for a human being."

⁓

The oppressive heat and humidity had draped us like a wet shroud during our visit in Nara, the ancient capital of Japan. Accustomed to the dry heat of Colorado summers, my family could venture out only a few hours each day before retreating, descending to the coolness of our garden-level apartment. During the summer of 1994, a few years after our work together in Hokkaido, Nakano had found funding to invite me back to Japan to visit and speak at ten different universities and national laboratories where scientists were studying freshwater fish. Having missed my family greatly during our five-week expeditions, I arranged for my wife Debbie and our eight- and eleven-year-old children to accompany me for the first ten days of this trip. We were happy to board the air-conditioned train in the heat of Nagoya for the trip north into the Hida Mountains of Gifu Prefecture, neighbor to Nagano where the 1998 Winter Olympics were held a few years later. We had seen beautiful temples near Nara and Kyoto, and visited Mount Hie and seen the Japanese monkeys that Kyoto University primatologists had studied. Now we were ready to find cooler temperatures at higher altitudes away from the coastal plain.

Nakano had arranged for us to travel with his family around his home region in Gifu and see the beautiful mountains and riverscapes that he had explored as a youth. He also wanted me to witness firsthand some of the rivers in which he had snorkeled to study the social behavior of charr and salmon before going north to work in Hokkaido. Views at every turn were rich and exotic, graced by aquamarine rivers flowing through deep, bouldery canyons. Verdant forests of beech and maple, fed by frequent rains, enveloped the mountains beyond, in ways reminding me of the Appalachian Mountains and their deciduous flora. Only much later did I learn that Nakano had returned to his home town after college to explore with a colleague deep into the mountain headwaters of these rivers and conduct some of the first sampling of fishes to determine where different species occurred throughout these remote reaches.

As we traveled, my family found wonder in the culture and people of Japan, attempting like all Americans to reconcile traditional houses and dress and food with the pervasive influence of our Western culture on everything from fashion to architecture. Shigeru's gracious wife, Hiromi, was a welcoming host for my wife and children, all the while caring for their own two-year old daughter Nana (and expecting another child). Hiromi had been an airline attendant within Japan, and so spoke English rather well, although haltingly. She and Shigeru had grown up only a half a block apart and had met in junior high school. They married when Shigeru returned home after college. Hiromi took great delight in becoming acquainted with Emily and Ben and did all she could to help them feel comfortable so far from home. But even though we had traveled extensively with our children in North America, they weren't prepared for such a drastic change in culture and food.

The beautiful dinner lay exquisitely arranged in intricate detail on the low tables in the traditional *ryokan* the first night of our visit. Each of the small flowered dishes held a carefully prepared course of miso soup with seaweed, tofu, cooked wild vegetables, or unfamiliar-looking broiled and raw seafood. Although I was accustomed to eating all these, only then did I realize that none of them contained foods that were familiar to my children, except perhaps for the beautiful covered bowl of white sticky rice. They looked astonished as they surveyed the meal. Peering into one of the small bowls, their looks changed to revulsion as three pairs of vacant eyes stared back at them, belonging to three small cold boiled squid, pure white with purple speckled backs. Poking at their food throughout dinner, they ate mainly rice.

On the third day, we reached Kamioka, the small mining town settled into the beautiful Takahara River valley where Shigeru and Hiromi grew up. Shigeru had arranged for us to stay the night in his parents' home, a wonderful hospitality to us as Americans. Nestled among the other dwellings along the curving narrow village street, their house is sided with rough-sawn dark cedar, and a pine tree grows in the small alcove at the entryway. Trunks of large cedars form the corner posts in the front, supporting the roof of clay tiles. The beautiful home is about two hundred years old and very traditional, with many small passageways connecting different rooms on two floors. A tiny garden and pond in the center open to the sky, welcoming the light and offering views of nature and fish. While in high school, Shigeru had replaced the traditional fancy *koi* in the pond with native charr from a local stream. Floors in the main living rooms are covered with beautiful *tatami* mats made of woven soft rush-es, and the sliding rice paper walls are bordered with fine-grained wood coated with traditional *urushi* lacquer, a rich dark butterscotch in color.

I realized later that Nakano could see the difficulty our children were having with the Japanese food and their unfamiliar surroundings. After we were introduced to his parents and invited into their home, his mother and aunt disappeared with Emily in tow, who, at age eleven, was the same height as they were. They led her off through the warrens of their home into another room, and later she appeared wearing a beautiful *yukata* robe, the traditional summer apparel for Japanese. But the dinner his parents prepared was one that I will always remember. Nakano's mother and aunt ran a banquet restaurant in their home, which had a large room upstairs to entertain guests. When we arrived there for dinner and knelt at the long row of low tables, we soon were served platters of mouth-watering beef steak, and fried pork and fish, the best I have ever tasted. His father then brought more trays with white bread, yogurt, milk, and ice cream—foods much less common in Japan. We all ate until we were completely full. I could find no words to express my gratitude for their consideration of my family. This experience in 1994 created a bond of friendship with Nakano and his wife and parents that has lasted to this day.

∽

After the accident, my memories of the rest of spring semester in 2000, and throughout the summer and early fall, are of feelings of deep, profound loss. I thought of his family who had lost their husband and son, so vibrant and rich with life at its prime, and one who had propelled himself from humble beginnings in Kamioka to become an internationally recognized ecologist. More importantly, he left three small children, then only one, five, and seven years old, now bereft of a father. I also worried about my colleagues in Japan, who had lost three of their best ecologists. Takuya Abe was a world expert on tropical termites. Masahiko Higashi was a respected scientist specializing in ecological theories and predictive models, who had worked closely with American ecologists. And all three of them left a network of students and close collaborators, among whom they had earned great respect and shared deep friendships.

As for myself, I had also lost one of my closest colleagues and best friends. Nakano was the one who really understood my research, and my passion for it, the best. But others wondered why I was so strongly affected. After all, here was a person who I visited only sporadically every few years, and who spoke English only with difficulty. The graduate students in our laboratory most appreciated what his loss meant. The support they offered provided me the greatest solace, even though I knew that others needed more comforting than I.

Nakano's wife and parents returned home after witnessing the search for two weeks in Mexico, without any sign of their husband and son, and no opportunity to reach closure. I wrote them letters, offering the deepest sympathy I could describe, and explaining how other scientists around the world like me remained in awe of the amazing achievements their son had accomplished. They struggled during the summer and early fall to hold two funerals without his body to bury, one in Kamioka and the other in Hokkaido where his wife and family live. Dull depression edged my thoughts throughout, and I found relief only in the hard labor of field research in the mountains of Colorado that summer. I must have talked about the accident incessantly to others, but I honestly don't remember.

I do remember thinking, "Should I close the book on Japan?" Would it be best to simply end that chapter of my life and career and consider it a wonderful, bittersweet memory? Or should I try to go back? I turned the idea over in my mind for many months, trying to view it from every angle, during the fall of that millennium year 2000. Could I somehow find a way to continue Nakano's legacy and follow the clues that he left us? "But I am a fish biologist," I thought, "not an ecologist like him, who had studied all the parts of the food web!" What did I know about stream invertebrates and algae, much less terrestrial insects and spiders and birds? The entire topic seemed daunting, and yet also somehow critically important.

Another strong feeling began growing somewhere in my subconscious and soon started to overwhelm these doubts. I began to realize that I wanted to find a way to help the graduate students and young colleagues that Nakano had mentored. I had worked very closely with Satoshi Kitano and Yoshi Taniguchi and had come to know many others during our research and travels in Japan. Everyone needs someone to write to with questions about how to negotiate the uncharted waters of their scientific career, to offer advice and encouragement through hardships and failures. I wondered if I could offer that kind of support for the young Japanese scientists like Yoshi and Kitano, and perhaps some of the others who Nakano had worked with more recently. Finally, I decided to write Yoshi and Masashi Murakami, Nakano's most recent postdoctoral research scientist, who had just been hired as an assistant professor at Tomakomai Experimental Forest to replace him. Did they want to collaborate, and try again?

The still, cold winter air held the car exhaust in a pall close to the ground as I hiked along the busy city streets from the university apartments to the dingy cement building that housed the Center for Ecological Research at Kyoto University. In early January, doorways in Japan are decorated with *kadomatsu*, garlands of green pine boughs, bamboo, and red or orange berries or flowers bound in straw rope, to welcome the ancestral spirits back and ensure a bountiful harvest. This month also includes Coming of Age Day, when twenty-year-olds celebrate adulthood. By this time in January 2000, Nakano had moved to the CER, as it is known, and settled into his new quarters. He was busy developing new plans for research near Lake Baikal in Russia, while finishing other projects in Hokkaido and Borneo.

I had traveled to Japan during my short winter break to work on two papers that Yoshi and Nakano and I wanted to finish—finishing being the hardest part of any research. Nakano and I both knew that this kind of face-to-face retreat was the surest and most efficient way to actually complete manuscripts. He regularly lured his former students like Yoshi back to his office for a *ronbun gasshuku*, a "paper retreat," where they worked nearly continuously, through day and night, to reach the goal. Unfortunately, Nakano became sick during my visit and took several days to recover.

But during my visit, Yoshi also wanted to celebrate his good fortune and the fruits of their hard work. How improbable it had been for him, the son of a *koi* farmer (more correctly, *nishiki-goi*, or ornamental carp), to travel to the United States to earn a second college degree, meet me, later earn his master's degree at another university, and then have the chance to work with Nakano on charr in Hokkaido. After completing and publishing all this research, how unlikely it was to have then earned a faculty position in a Japanese university, a career that often eluded even the best young scientists in Japan. We traveled by train to Matsusaka, southwest of Nagoya, where his family home lies surrounded by rows and blocks of his father's fish culture ponds. Sitting in their sunroom with Yoshi as interpreter, I learned that the largest and fanciest fish with certain unique color patterns could be sold to wealthy foreigners for more than $10,000 each.

In honor of my visit, Yoshi's parents sent us to a beautiful hotel near Toba, along the coast of Ise Bay, about an hour distant. Here the low but rugged mountains of the Kii Peninsula, covered with a billowing mantle of rich green subtropical forest, kneel down to the sea and lend a striking backdrop to the long traditions of pearl culture and diving by Japanese women to harvest abalone and other foods from the sea. The other guests at our hotel were mainly twenty-year-olds, focused on a boisterous celebration of their new-found freedom as adults. Our dinner, served in the room after the traditional evening bath in the hotel's hot springs, was among the most amazing meals I have eaten in Japan. Arrayed on a large wooden platter, it was almost entirely fresh seafood, with many kinds of *sashimi* (extremely fresh raw fish), including puffer fish (*fugu*). Knowing that this could contain a deadly neurotoxin if not properly prepared, I decided to boil mine in the water reserved for boiling vegetables. I'm not sure it would have made a difference.

During my visit that January, Nakano and I also wanted to discuss how we might collaborate on research again. A few years before, the grant proposal we had written to the same joint American-Japanese program that had funded our first work had failed. By then, the goals of the two agencies seemed just too different to allow one proposal to earn high scores from both. As Nakano recovered from his bad cold, we wandered the streets near Kyoto University, visiting teahouses and discussing many ideas. He told me of his findings from the Hokkaido study with Masashi Murakami, about how streams and forests supplied food to each other across the aquatic-terrestrial boundary, and how this supply alternated across the seasons. Intrigued by these unique results, I encouraged him to submit the paper to one of the premier scientific journals in the world, which he and Murakami did in the next few months.

For future research, Nakano was most interested in integrating humans and the natural environment together, and he had wide-ranging ideas. We talked of studying how human actions, such as deforestation, could disrupt the subsidies between streams and riparian forests. Given our difficulties with the previous proposal, he also suggested that we collaborate informally. I would write a grant to the National Science Foundation alone, and he would write separate grants to the Japan Society for the Promotion of Science to support postdoctoral researchers (also known as postdocs) who could collaborate with my group. After a week filled with intense interactions on manuscripts and proposals to advance our science, and celebration of Yoshi's accomplishments and good fortune, I stepped outside the CER to say good-bye and catch my ride to the train for the airport. Although I'm usually not sentimental about saying final farewells at the end of a visit, on that day I decided to take a picture of Nakano and his students to remember our time together. I didn't realize that those would be the last moments I would see him on this earth.

∽

The NSF grant proposal deadline for the collaboration I had pondered loomed in early December 2000, only a few weeks away. By the time Yoshi and Masashi Murakami wrote back that they were interested in collaborating on new research in Japan, it was nearly Thanksgiving, when the pressures of the fall semester weigh on every waking hour. How could I ever finish a proposal by that grant deadline? Most scientists will tell you that writing an NSF grant proposal that is good enough to earn funding is as hard as writing a paper that is good enough to be published. Even when you have good ideas, and they are all in order, it can take an entire month. I had only two weeks.

But a deep longing, a passion to pursue my friend's ideas and keep them alive, made me determined to try. As I started into the work, I soon found myself living on the edge, doing only what was absolutely necessary to teach my course and help my graduate students, and missing many other deadlines. I stole time from my family and sleeping to read, think, and write. Most meals were eaten at my desk, in front of the luminous computer screen. I knew the main results of Nakano's work on stream-forest connections, but I had not read those papers in detail. I had taken courses in college on aquatic insects but had never conducted research on these invertebrates or their emergence. I knew nothing about stream algae, or spiders. But I also knew that I was interested in even more than the intricate ecology of these relationships that Nakano and his colleagues had studied previously. I wanted to explore uncharted territory in the ecology of streams and forests. Drawing from the ideas and experiences that Nakano and I had woven together the last time I saw him, I became interested in how humans could affect these webs of food that move across the stream-forest boundary.

And as I read deeper, poring over the papers he had written with his colleagues, the clues for a next study nearly leapt from the pages and arranged themselves in logical order in my notebook. Cutting off inputs of terrestrial insects to Horonai Stream using a greenhouse had huge effects on algae because the Dolly Varden charr were forced to shift their feeding to bottom-dwelling invertebrate grazers. This shift happened very quickly, just as we had found in Poroshiri Stream when we filtered out the drifting invertebrates. But in another paper their data also showed that nonnative rainbow trout usurped most of the terrestrial insects that fell into Horonai Stream. Could this nonnative trout have effects similar to those of cutting off the inputs using a greenhouse? Would the effects of either the greenhouse or the nonnative trout then also cascade through the stream food web, causing Dolly Varden to reduce aquatic insects, in turn leading to reduced emergence of adult insects that feed spiders in the riparian zone? Then a larger question slowly emerged, one that could offer the key reason that scientists on the NSF review panel might consider the research important enough to fund. Could studying these two factors in a large-scale field experiment capture the essence of the most common effects that we humans have on streams all around the world? We destroy riparian habitat, often by deforestation, and introduce nonnative species like rainbow trout.

I worked steadily, doggedly, stealing time every day and night, and all day on weekends. Every possible waking hour was absorbed in reading, writing, planning, and budgeting for a research project that would be half a world away. It would be among the largest-scale experiments of its type ever conducted in a stream, using sixteen fenced sections laid out over more than a mile of the channel. I made careful predictions about the effects on the fish, aquatic invertebrates, stream algae, and emergence of aquatic insects of four different experimental treatments that we would create in the sections. Compared with when Dolly Varden charr were alone, I predicted that either adding the greenhouse or adding rainbow trout (or both) would reduce the supply of terrestrial insects available for the charr to eat. Based on our earlier work, I reasoned that the Dolly Varden would then quickly shift to eating more bottom-dwelling stream insects, and reduce their numbers drastically. In a cascading fashion, this would then cause an increase in algae as these grazers declined, I thought, but also shut off the supply of adult aquatic insects emerging from the stream.

As words for the proposal flowed from my mind through fingers and keyboard, I felt Nakano's spirit, somehow spurring me on. Not only were all the clues for this next step in his research lying just beneath the surface of the pages in his group's most recent papers, but the infrastructure that he had carefully created at the Tomakomai field station stood ready to support it. The logistics of this kind of large-scale field research are daunting. The list is nearly endless about what is needed to conduct the work, and field studies wear out or destroy equipment on a regular basis. I needed greenhouse frames and the mesh covers, plastic mesh fences to enclose the fish, crews to assemble both of these, and trucks to haul gear and sandbags to the stream. We would need electrofishing equipment to sample fish, and pan traps and emergence traps and drift nets and many small pieces of gear to sample invertebrates and algae and fish stomach contents and spiders. We needed laboratory equipment like microscopes and drying ovens and balances for sorting and weighing all the samples. And even more than these, I needed technicians, not only to conduct the field research, but to sort and process all the fish diets, invertebrates, and algae samples.

But the amazing thing was that nearly all of this was available, created by Shigeru Nakano when he was Director of the Tomakomai Experimental Forest and a scientist there. The apparatus, the equipment, the crews to assemble them, the other graduate students who could help, and even the local women whom he had trained to identify and sort stream and forest invertebrates, were all there! Many of the people were working on other projects headed by Murakami and other staff, which Nakano had helped to start. And all these people and equipment were within about 2,000 yards of the field site on Horonai Stream where we would work, a major advantage compared with any other field research I've done. The field station also had dormitories for us to stay in, and even provided meals for a modest fee.

At three o'clock in the morning on the day that the proposal was due, I had finally finished typing the last forms. By noon I had survived the bureaucracy of budget negotiations and required signatures, made more difficult because I became sick after so many days with so little sleep. But would the panel at NSF support the proposal? I didn't think so. I told myself that this was just the first step. Success rates were so low that the proposal would surely be rejected. Nevertheless, I was determined to take their comments and revise it for the next opportunity, probably a year later. In the end, a colleague on that panel later told me that it was one of the best proposals they had seen, and they funded it on my first attempt, for the entire budget (also rare). Only about one in seven proposals was accepted and funded on the first try, and several of my proposals had been rejected in the previous five years. This success was as far from what I had expected as Japan is from North America. Somewhere, that day I heard the news, Nakano was smiling.

‹ ›

"Fausch-san, can you help me with my manuscript?" asked Yoichi Kawaguchi, one of Nakano's last PhD students. It was about 10 p.m., and I had spent the day wading up several miles of Horonai Stream to plan the research we would conduct a year later. When I learned in May 2001 that my grant would be funded by NSF, I was able to return to Japan in June after my classes ended to arrange the many things that would be needed for such a large-scale research project, so far away. There were permits to apply for, and housing to arrange, and a vehicle to buy for the postdoctoral researcher I planned to hire. I also wanted to inspect the actual greenhouse apparatus and other equipment that we would use for the research. But I did not realize that my visit would also become a kind of reunion for Nakano's former students and colleagues, many of whom returned to Tomakomai to rekindle our friendships and remember his influence on our work and lives. Satoshi Kitano came from Nagano, where he was working for a regional research institute for environmental conservation. Yoshi came from his university in southwestern Japan, and

Kawaguchi from his postdoctoral research position in Gifu Prefecture in central Japan. A year after the accident, we all felt a kinship in supporting each other and an urge to move as best we could from daily touching our keen sense of loss toward seasons of finding ways to honor his legacy and be grateful for his life, and ours.

"They found my manuscript in Nakano's bag in Mexico, but he had edited only about a third of it. I need help preparing the rest for publication," said Kawaguchi. Yoshi offered, "I can also help interpret for 'Gucchi', if you want." We had finished an impromptu dinner of Japanese take-out food in the Tomakomai dormitory, where we were all staying, and had enjoyed drinking beer and savoring our finest memories of Nakano for several hours. I didn't think I had the energy to focus my mind on revising a manuscript, especially one of such importance to a former PhD student of my colleague. They had hoped to submit it to *Ecology*, the premier ecological journal in North America. But my visit was only for a few days, and the look in Kawaguchi's and Yoshi's eyes told me that I needed to try. I wanted to help Nakano's former students, and here was an opportunity for a *ronbun gasshuku* that I couldn't pass up.

Sentence by sentence, we worked through the manuscript from start to finish, turning unclear statements and much "Japanese English" into scientific writing that I thought might be acceptable to the scientists who would review the work for *Ecology*. Half-heard discussions by our colleagues in the halls of the research station gradually died away as they filtered off to sleep. Cicadas buzzed loudly outside in the warm humid darkness, their sounds rising every few minutes, then falling away. By about 2 a.m. we had finished making the changes. The results of the experiment by Kawaguchi, Yoshi, and Nakano showed that when they installed greenhouses over stream reaches and fish were allowed to move freely, half the weight of charr and rainbow trout left the sections. Their landmark study showed that when the terrestrial insects that supply half of the fish's food were excluded, half the fish left. The equation was simple, and the work is among the many elegant research accomplishments by Nakano and his students. For me, his spirit walked the hallways and visited the rooms of the Tomakomai research station that night, and has done so each time I have returned.

The next day I was happy when Nakano's wife, Hiromi, visited the station, although she appeared still rather shocked and depressed from what she had been through since March 2000. She also looked different than I had remembered her from 1994, when our families had traveled together to their hometown in the Hida Mountains. As we visited over dinner with Nakano's former students and colleagues and played with their children at their home nearby, I talked to her about the accident and her feelings about trying to make sense of it all. I gently encouraged her to move forward, to think about what could be done that would lead to good things happening in the future for herself and her children. Although focusing on positive things can never heal all wounds or relieve all the feelings of loss, Debbie and I had found it the only way to move forward and help Emily and ourselves after her difficult birth. To do otherwise only makes a person less able to embrace the truth that life is good, by definition, and to help the others around you. By the next time we met, a year later in 2002, Hiromi seemed a bit happier and calmer. Each year put distance between her and the memories of those weeks. Like an oak tree in the garden, some of life's most necessary gifts require nurturing, and grow slowly. I can only hope that my words helped her and Nakano's colleagues by some small measure.

I try to draw lessons from the tragic loss of the scientists and a good friend, after time and circumstance have ordered these events among my own life experiences. What was there to be learned? What goodness could I carry with me, earned through the grief and renewal? Certainly, one lesson is that good friends and close colleagues, as good and close as Shigeru Nakano, are very rare. I've found only a few colleagues since then that have the innate understanding he had of my ideas and values in science, and what has driven me to explore them. In return I certainly understood his passion, and saw his promise, and others tell me that I helped nurture his ideas and influenced his approach to research. Such close connections are to be cherished, for they may not occur even once in the life of a scientist.

A second lesson is that great mentors inspire. Even one statement or action can create a memory that endures forever in the minds of those seeking a similar pathway. When Nakano's former students were interviewed for a documentary film in 2005 and again in 2012, many talked about what had become known as the "Nakano School," in which the close-knit group strove so hard toward a common goal at the Tomakomai

Experimental Forest and beyond. Nakano's drive to do the best research humanly possible, coupled with his passion for nature and his need to understand it, left an indelible mark on everyone with whom he worked, including his students and me.

On the day before he left California for Mexico, Nakano called Daisuke Kishi, another of his PhD students, who was working on a new experiment at Tomakomai. They had planned to test whether changes in water temperature, such as those predicted to be caused by climate change, would alter the feeding relationships among Dolly Varden, benthic insects, and algae. And, they wondered, could these effects be strong enough to fundamentally change the trophic cascade, whereby fish increased algae by eating grazing insects?

"My experiment wasn't going well," Kishi recalled later, "But, in his last words to me, Nakano-san urged me to work hard on it." He used the words *Gambareyo na!* which mean "You must do whatever it takes!" Kishi felt dejected, because his experiment was turning out badly, and said that he felt like he should quit, yet continued anyway. Unfortunately, the experiment did fail, and he thought Nakano would be mad at him. But then he heard about the accident in Baja.

In Kishi's words, he later thought, "I can't let it end like that. So, I did the experiment again." He wrote, "It took me three years to get it right and analyze the samples and data. But, I finally did all of that, and the paper was later accepted. I just wanted to let Nakano know that I completed the research that we planned, and it was accepted by other scientists. I just wanted to show him how much I appreciated everything he did for me, and to see him happy about that."

In the end, Mary Power, the world's expert on trophic cascades in streams, highlighted Kishi's experiment as the first ever on this critical topic of how changes in temperature can disrupt food webs in streams. Respect and praise by our fellow scientists, especially those in the best position to know, is the highest honor that any of us can imagine. It strikes me that, like being a parent, teaching and mentoring students is a long-term study in delayed gratification. It can take many years, perhaps even a decade, to see their work, and your work with them, come to fruition. It may take more years before that work is recognized and used by other scientists. The research by Nakano and his students is the kind of science that others use, and the results are now well known throughout the world. I hope that somewhere Nakano can enjoy the triumph of seeing the many successes by the students that he mentored and inspired.

My experiences after the accident were clouded in my own grief, and the busy press of academic work and helping raise children as life went on afterward. I'm sure it was far more difficult for his family and his colleagues. It was only much later that I learned more about what it meant to my Japanese colleagues for us to return to conduct new research with them and to create new collaborations. Several said in interviews how much it meant to know that I would come to Japan again and help "do good research" that would further Nakano's ideas. For my part, I also realized just how much I had enjoyed visiting and working in Japan, and the rich tapestry of people and culture that drew me back.

But the clues that Nakano left in his papers were the flame that inspired me the most, and showed just how brilliant a scientist and exceptional a mentor he was to all of us. Several of his students spoke about how he could often see both the problem and solution very quickly, even after only a few days working in the field. And I knew that he then had the drive to work for months and sometimes years to gather the data needed to test his ideas. I wanted to go back and follow up on the clues he was pointing toward, and write another chapter in Japan.

Chapter 5
Riverwebs

"*Iki ma sho!*" (Let's go!), I cry urgently to the Japanese forest workers and graduate students as the truck, loaded high with steel-pole greenhouse frames and huge bolts of fine mosquito netting, lurches to a stop along the dirt road that parallels Horonai Stream. The light green fern fiddleheads of spring have pushed through the matted brown plant litter on the forest floor and the sun shines brightly on the stream through the bare branches of maples just leafing out here in the last few days of May 2002 in southwestern Hokkaido. Our crew of Japanese and a few Americans sets to work carrying the supplies through the riparian forest to the stream, assembling the broad U-shaped frames and pushing their points into the soft gravel along the stream banks so they stand upright. When the poles that make the side and top rails are attached, our creation becomes a large segmented caterpillar arching over the stream, about 14 feet wide and 7 feet tall, snaking 60 yards through the forest.

A hundred yards upstream, Colden Baxter, now a postdoctoral research scientist in my laboratory, is leading another crew that has already started covering another greenhouse frame with insect netting. They unroll the sturdy fine nylon mesh from bolts mounted on the back of the truck and lead it through the maze of trees, pulling it hand-over-hand, one person to the next. It takes about one bolt for each side panel, and one joining them across the top, to cover the whole frame. They secure the mesh with plastic clips, stake it to the ground on each side, and drape it over the upstream and downstream openings to completely cover the reach of stream.

Very soon, hundreds of adult mayflies and caddisflies emerging from the surfaces of pools fly up and congregate just beneath the crest of our stream veil, their translucent wings catching the light as they bustle upside down along this corridor. Most eventually make their way to the ends and escape through vents we cut there to release them. Forest caterpillars of all stripes, and several kinds of ants, drop from the trees onto the greenhouse and crawl across and down the sides, lucky to have avoided the fate of falling into the stream and being eaten by fish. Shiny black ground beetles wandering incessantly across the forest floor meet the mesh wall and must detour, even though some would have tumbled from the banks into the stream. The fish in the stream and the spiders along the banks begin to consider their options, as the prey they had become accustomed to eating grow scarce.

Bone tired after many days of working throughout the stream, we pack up our tools in the lowering rays of late afternoon sun, pausing briefly to marvel at the intense spring green of new leaves as they are backlit from the west. We pull off the waders that usually keep water out but also hold the inevitable perspiration in. The extra tugs to remove soggy socks play on our fatigue, but make dry shoes a welcome reward at the end of the day. Just then, a thought overtakes me, penetrating my weariness. Just like starting the first field experiments that Nakano and I did together in Poroshiri Stream in 1991, it felt good to finally be *here*, doing the work that I had first envisioned eighteen months earlier. We had finished two greenhouses today, but it will take at least three more days to build all eight we need for one of the largest field experiments ever conducted in, and over, a stream. And, it felt good to come back to Japan and work with Nakano's former graduate students Masashi Murakami, Yo Miyake, Yoshi Taniguchi, Mikio Inoue, Daisuke Kishi, and several others on this ambitious new research. Our recent shared loss made this chance to honor Nakano's legacy by doing new cutting-edge research all the more meaningful.

How *do* the things we humans do to forests, and to streams, affect these ecosystems that are joined across their boundaries? Can we as scientists predict what will happen if humans cut down the riparian forest, or simply kill the insects there, such as by spraying them? What about when we stock nonnative trout that we like to catch into streams, or when they escape from a fishing pond during a flood and invade on their own? These two types of actions are among the most common effects that humans have on streams, and they often occur together in the same places. From earlier research by Nakano and his students, we knew that cutting off the "subsidy" of insects falling into streams, such as would occur when the forest is removed, has big effects on the stream ecosystem, and that these effects can cascade to the very base of the food web. But would the effect of introducing trout be similar? And, if these actions caused fish to reduce the aquatic insects in the stream, would this, in turn, reduce the emerging adult insects that feed animals such as spiders and bats in the riparian forest? Nakano's research had fueled our desire to know the answers, and we were ready to stake our futures on a massive and grueling field experiment to find out.

In the real world where we humans affect streams, many other things are happening at the same time, and everything is connected to everything else. Let's say, for example, that we want to measure the effect of nonnative rainbow trout that have been introduced for angling and are now invading many streams throughout a region. However, in one stream water is diverted by a city using a small dam, so the natural flow is altered. Near another, agricultural fields are sprayed with a chemical to kill insect pests, but that insecticide washes into the stream after rains and also harms fish and aquatic insects. With all these other things happening, it is very difficult to determine whether introducing rainbow trout has a consistent effect or not. We need an experiment to find out.

Experiments are the best way that scientists have invented to figure out the effects of one factor at a time, by keeping others things as constant as they can. This is most easily done in a laboratory, where most conditions can be controlled. For example, if scientists want to know the effects of the insecticide described above on fish, they can compare fish that are exposed to the chemical in tanks in the lab (called the "treatment" fish) with fish in other tanks that are not exposed (the "control" fish).

Changing only one factor at a time (such as the toxic chemical) while keeping everything else the same (such as using fish of the same size and keeping the water at the same temperature) is one of the three hallmarks of a true experiment. This method prevents other factors from affecting the results. For example, if all the fish used in the experiment are captured and handled the same way, then the stress of this handling should affect the results for both treatment and control fish the same way. Therefore, any difference in survival must logically be caused by the one factor that is different, namely adding the chemical. In short, if the experiment is "controlled" so that "all other things are equal," then this allows scientists to measure the true effect of the one factor that is different, in this case the insecticide.

The idea of consistency or repeatability is the second hallmark of experiments. For example, what if we compared only one treatment fish in a tank of water with the insecticide to one control fish in an identical tank but with clean water? Then, there is a pretty good chance that some characteristic of one of these fish would cause it to survive better or worse than the average for other fish held under these same conditions. We are likely to get an inaccurate answer about the effect of the chemical, or even the wrong answer, if one fish is odd. Scientists want sufficient evidence that the insecticide consistently reduces survival of the fish, so they test many fish in separate tanks with and without the chemical. This is called "replication." The more consistent the results are when many fish are tested independently, the more confident we are that we understand the true effect of the chemical and can use these results to predict the effects in nature.

The third hallmark of a true experiment is that fish must be randomly assigned to be either exposed to the insecticide or not, to ensure a fair test and avoid bias. If, for example, a technician setting up the experiment feels sorry for some fish that are slightly smaller or less robust, and chooses fewer of these to be exposed to the chemical, then the results could be biased. To avoid this, scientists go to great lengths to ensure that fish are chosen randomly (or "blindly") to be treatment or control fish. This process is called "randomization." To sum up, control, replication, and randomization are the three elements that make up the gold standard by which scientists judge whether an experiment is designed correctly so that the results will show the true effects of the factor tested.

A fourth characteristic that is critical in ecological field experiments, but very difficult to achieve, is "realism." If the question a scientist wants to answer is simply how much insecticide will kill fish, then the laboratory experiment above can provide an answer. However, in our case, we wanted to know whether Dolly Varden would affect an entire stream-riparian food web differently when nonnative rainbow trout invaded (that is, were added), or when the stream was covered with a greenhouse that mimicked deforestation by preventing most terrestrial insects from falling in. So we needed to create "experimental units" that were not individual fish in tanks, but entire 30-yard reaches of stream with fences on each end to enclose groups of fish and the invertebrates and algae that create the food web, complete with the adjoining intact riparian forest. Given this, how could we set up a realistic experiment in Horonai Stream and still meet the gold standard of control, replication, and randomization?

For all our reaches, we wanted first to mimic the natural situation, with the same number of native Dolly Varden charr found in other Hokkaido streams, on average. This would serve as our control, against which we could compare three different treatments. Data from our previous work in Poroshiri Stream and other streams in Hokkaido showed that we needed forty charr to achieve this natural density in each 30-yard reach. Then, we could randomly choose some sections to be covered by greenhouses (one treatment), and others to receive rainbow trout at the densities (twenty fish per reach) that Nakano had measured in Hokkaido streams where they had invaded (a second treatment). Lastly, we randomly chose some sections to receive both the greenhouse and rainbow trout, a third treatment. To determine the consistency of the response, we calculated that we would need four replicate reaches with each of the four conditions (control and three treatments), for a total of sixteen study reaches.

But there's more. Water flows downhill, and so downstream. This means that the effects of a treatment in one reach could affect another reach just downstream that received a different treatment. For example, invertebrates from a control reach with only Dolly Varden could drift into a treatment reach where rainbow trout had been added and bias the results measured there. We expected there to be more aquatic insects with Dolly Varden alone than when rainbow trout were added, so if the reaches were too close together our experiment might yield results that were not accurate. Fortunately, ecologists who study aquatic insects have discovered that most species drift only a few yards before they land and take up residence again. To avoid this problem, we laid out our study reaches to be about 90 yards apart, on average, which is many times farther than most invertebrates drift. In addition, we surmised that fish living in these buffers between the fenced reaches would eat the drifting insects and prevent most of those leaving upstream reaches from drifting into downstream reaches.

What about randomization? Colden Baxter and I spent months during the winter beforehand planning the experiment, including how to lay out the sections in Horonai Stream based on the measurements I had made in summer 2001. We knew that if we simply assigned the treatments randomly to all sixteen reaches, there was a good chance that several reaches with one particular treatment (say, the one with greenhouses) could end up near the downstream end where the stream was slightly wider. Several of another treatment might then end up near the upstream end, and all of this could lead to biased results. So, to avoid this, we divided the stream segment into four equal "blocks," and laid out four study reaches in each block. Then we assigned one of each treatment at random to the four reaches in each block. This is a common method used in experiments like this, where another factor like width changes in a systematic way across the whole study area. It ensures that the four replicates of each treatment are laid out randomly, but are also arranged across this upstream-downstream gradient of stream width, so the results are not biased.

All our careful planning looked great on paper—but how difficult would it be to construct all the apparatus along Horonai Stream and to actually *do* the study once everything was built? Our most persistent memories of setting up this huge experiment are of rain and sandbags. Soft gentle *kosame* (light rain), growing sometimes to steady rain, created a fine mist that seeped under our raincoats and down our waders and mixed with the sweat from hard labor. By the end of the day we wondered why we wore a raincoat at all. We spent long days constructing the thirty-two fences we needed to enclose the sixteen reaches, each supported by a framework of sturdy pipes, and more days building the eight greenhouses needed to cover the randomly selected reaches. Fortunately, Nakano had done all of this before, so all the tools were available and the methods worked out. And, as the enormity of the task revealed itself, we recruited more and more help, from the forest workers at the Tomakomai research station, from our colleagues, and from the many graduate students working on their own ecological research projects.

None of the forest workers spoke any English, so our Japanese colleagues and their graduate students who knew some of our language were pressed into service to translate. Colden had also hired Jess Jordan, a student from Oregon State University (where Colden had earned his doctorate), who spoke fluent Japanese. When we gathered at the stream the first day, Jess greeted the forest workers in Japanese. "*Ohayoh gozaimasu!*" (the most polite form of "Good morning!" in Japanese) he began, emphasizing the last syllable of each word strongly. "*Ohhh, Osaka-ben!*" exclaimed several of the forest workers, recoiling in great surprise over his thick regional accent. Jess had worked for several years in the Kansai region near the city of Osaka, a region with a dialect that is both melodic and harsh. Even someone like me from the West can easily tell the difference. In comparison, the forest workers had all grown up in rural Hokkaido and spoke in gentler tones. It was as though a Brooklyn native had traveled to northern Minnesota.

Horonai Stream traverses a flat plain covered by pumice gravel that rained down most recently when the Mount Tarumae volcano erupted about three hundred years ago. This is the most active of the three volcanoes that rim a collapsed caldera formed by a much older and larger volcanic eruption. The depression filled with water, forming Lake Shikotsu, a very deep "crater lake" about 15 miles west of the Tomakomai Experimental Forest. This bed of pumice creates a porous aquifer filled with groundwater from rain and melted snow, which seeps into the stream and provides an extremely stable flow with few floods. However, the pumice gravel is also so lightweight that when you stand in one place, the flow swirling around you can wash the gravel away and dig a hole around your feet.

This pumice created a big logistical problem. How could we build fences on such an unstable streambed? Any small current around them would wash away the lightweight gravel, creating a gap at the bed or near the stream bank that would allow fish to escape. For the landmark experiment in which he first used greenhouses, Nakano had pioneered a solution, laying down plastic tarps and covering them completely with sandbags filled with more gravel. Each fence required about thirty sandbags just to create this stable base. We then pounded three or four pairs of pipes through the tarp and into the streambed. The pairs were arranged in a line across the stream and clamped together at their tops, one set upright and its pair angled upstream into the streambed. To the angled posts we then clamped a framework of three horizontal pipes that spanned the width of the stream channel. Finally, we unrolled the sturdy plastic mesh and attached it onto the angled framework, and then cut, fit, and sealed it with more sandbags to conform to the contours of the streambed and banks. When complete, each fence was checked by snorkeling to make sure it was "fish tight."

And, as always in field experiments, we were racing against time. Colden had arrived two weeks before me, at the beginning of May 2002, and began the job of organizing and purchasing equipment and supplies and planning the work. But we knew that we had to get the experiment "in the water" as soon as possible if we hoped to have enough time for all the organisms to interact with each other so that we could answer the big questions we were asking. Populations of algae needed to be grazed by aquatic insects, and fish needed time to eat these larvae in order to alter emergence of their adults. The teeming masses of biota needed to fulfill their destiny over at least six weeks for us to have a chance of detecting changes caused by the different treatments.

On the morning of the eighth and final day of building fences, I arose very early and stepped out into the cold Hokkaido dawn. The light on the eastern horizon glowed spring green through the new leaves of maples as I started up the dirt road on a borrowed too-small bicycle to inspect and clean the fences we had constructed so far. Shreds of mist rose from pools where the stream meandered through a meadow, grass bright with dew, creating a peaceful scene. But after so many days of carrying sandbags through the forest and building fences in the cold rain, my body revolted and my muscles suddenly spasmed, gripping my lower back like a vise. Without warning, I was out of commission, and of no use in helping complete the fences. But Colden and Masashi Murakami and our Japanese colleagues continued on. Afterward, they set to the next task of removing all the fish from each of the sixteen reaches using electrofishing, so we could restock the numbers of Dolly Varden and rainbow trout we needed to create identical replicates of each of the four treatments for the experiment.

‹ ›

The key ingredient to the success of the entire field experiment, one of the most difficult and comprehensive ever done in a stream, was Colden Baxter. Raised on farms in Wisconsin and Montana, he excelled as a distance runner in high school. However, he fell ill with appendicitis during his senior year and so did not have the record of wins needed to earn the offer of an athletic scholarship for college. Despite this, he "walked on" and landed a spot on the varsity track team at the University of Oregon, the premier school for runners in the United States. I had picked the right person for the job. Only Colden could have run the marathon of our research that summer.

For his master's and PhD research, Colden had worked over great expanses of beautiful wilderness rivers in northwestern Montana and northeastern Oregon, his main mode of transportation simply hiking. For his doctoral dissertation, he wanted to know how the entire assemblage of fish species in the Wenaha River used habitat throughout the entire 30 miles of the mainstem, where shady reaches with scattered Douglas-fir were interspersed among open sunny expanses with lodgepole pine. Colden snorkeled during both day and night, and through all four seasons, to count and record the habitats fish used. He tracked other fish that he fitted with small radio transmitters, both on foot and from an aircraft. He and his wife (and assistant) Lenny hiked and camped along the wild canyons, through blistering summer afternoons and winter nights below zero, even during the first months of her pregnancy. Some tough couple.

Colden had completed majors in both biology and geology at the University of Oregon, which had endowed him with a nearly perfect blending of the big picture about how landscapes shape streams with the details of how stream organisms interact. During his master's work he had learned to identify aquatic insects and studied their ecology, and also understood how to sample and measure stream algae. In college he had taken two years of Japanese language, and could communicate with the forest workers. But most of all, his background working on farms gave him among the strongest work ethics I have known, and his athletic training gave him an equal sense of how to motivate a crew and "roll with the punches." It is an art to know when to press on, and when to quit, head for the showers, and go to town to buy the crew a good meal and a few beers. So when Colden showed up at the stream at 7 a.m. having already run 6 miles before breakfast, no one could claim that somehow they were being asked to work too hard. The example he set inspired others to dig deeper. His unique blend of intellect, experience, hard work, and good humor created the possibility of success.

The problem with working in streams is that they flow downhill. And in that flow they carry myriad pieces of living and dead vegetation from the riparian forest and stream, most of which catch on mesh fences. These pieces come in many different sizes, from branches and clumps of stream moss to last year's dead leaves partly chewed up by stream insects. As early summer progressed, we encountered an ever-changing procession of plant parts dropped by the forest into and alongside the stream. First came the "scales" that enclose the buds of large magnolia shrubs, followed by the flowers of maples and the catkins from birches, all clogging our fences. A bit later, windy nights brought in new bright green maple leaves that collected on the mesh and filled baskets as we removed them. When the debris built up, each fence became a small dam, forcing water to flow over, around, or underneath the matted plant parts, threatening to scour holes in the pumice streambed and banks. In addition, when the water backed up and spread out, it floated last year's dead leaves that littered the banks, bringing even more of them into the stream to clog the fences.

It soon became obvious that the thirty-two fences we had built were, as Colden observed, just like a herd of dairy cows. They needed to be tended twice a day, just like milking cows. Two people leapfrogged from fence to fence along the stream as they worked, brushing and scraping the wet leaves and sticks from each of the thirty-two fences with their hands so that the water could flow freely. Even using gloves, hands grew soft when constantly wet during the two-hour stint and were easily nicked and cut on the sharp edges of the stiff plastic mesh. Wind or rain brought in even more leaves and other plant parts, so on those days fences needed to be cleaned even more often. If we got too busy doing other work, or someone didn't do their "chores," the holes in the streambed and banks created by the incessant flow could take up to a dozen sandbags to fill and repair. Filling sandbags and carrying them to the stream, cursing intermittently along the way, reminded one that, like cows and small children, experiments in streams require babysitting 24/7 and can "melt down" if you ignore them. But by the time the experiment actually started, we had a sixth sense about when fences needed tending and had developed a system that worked.

We needed 640 adult Dolly Varden charr and 160 rainbow trout for our experiment, and the charr were not that common in the streams of southwest Hokkaido near Tomakomai. The best place to find them in abundance was Shiretoko Peninsula, the "last castle" that Ishigaki had written about, an eight-hour drive across the island to the northeast tip. Nakano had worked there before with his graduate students, and several knew the streams. So on May 30, with all the fences complete and the fish removed from each reach, we set out on an expedition to collect the fish. Masashi Murakami stayed behind and worked with others to install and adjust the final greenhouses and keep the fences clean.

Two of Nakano's most recent graduate students guided us on our journey and arranged the work parties. Yo Miyake, by then a postdoc, had worked for Nakano as an undergraduate on the first study using greenhouses and had stories to tell about the rigors of that work. He speaks

English well, and led the way. In contrast, Daisuke Kishi, a small, thin graduate student still working on his PhD research, spoke little English but was a dynamo in the field. We three Americans peppered him with questions in "fast English" as he drove us across Hokkaido in the research station's Toyota Land Cruiser, but after making a wrong turn he finally grew frustrated and declined to answer.

After spending the night in a small hotel in a seaside town, and a false start in one stream where fish were scarce, we finally began capturing the Dolly Varden we needed in a second stream. Yo and Kishi had warned us that brown bears are very abundant in Shiretoko National Park, so everyone was on full alert. This unfamiliar landscape was beautiful and yet beguiling to us as North Americans, like entering an Alice-in-Wonderland world where elements were put together that didn't belong. The stream cascaded down from the mountainous spine of Shiretoko Peninsula and through columns of huge boulders that formed pools up to a man's waist, treacherous to wade while carrying a backpack electrofisher. We had imagined a place like Alaska, where Dolly Varden are still abundant, with huge Sitka spruce and Douglas-fir shading the mossy-rimmed banks of pools. But instead of conifers, the stream was lined with a sunny, open gallery forest of scattered large old birch trees with trunks up to two feet in diameter. It was like no place we had ever seen where salmon or trout live in western North America. Brown bears and Dolly Varden seemed terribly out of place to us here, wandering among the birches.

Working quickly, and sooner than I had imagined, our two crews had the 700 adult Dolly Varden that we needed to be sure we had enough. I thought the Shiretoko streams would be unproductive, and that it might take much more work to catch this many fish. Kishi-san, quiet and stoic as ever, and despite his small frame, caught probably a third of the fish himself. The water was cold and deep around the slick boulders, and by the end nearly everyone had slipped at least once and filled their waders. Working with lots of fish near a stream is a dicey business, because passersby (including bears) always want to see your catch, or call the authorities. We moved the fish quickly from live cages in the stream to big plastic bags filled with water in coolers in our trucks. Even before removing our waders, we drove to another spot along the winding forest road in the national park to add compressed oxygen and check the fish, away from curious bears and people. When the fish were finally settled, Yo and I removed our wet waders and prepared for the long drive across Hokkaido. But when I turned to look at the others, there was Colden Baxter, stripped down to no clothes by the side of the road, hating the thick wet neoprene waders in which he had spent the day. Although we had permits to capture and transport the fish, I imagined a lot of explaining to park authorities if they happened upon this motley crew!

Dawn was creeping into the edge of the night sky by the time we arrived at the Tomakomai station at 3:30 a.m. and began settling the charr into aerated tanks in the fish culture outbuilding. From then on, we treated them with utmost care, checking the water flow and compressed air pumps every four hours through day and night. Colden worked with Yoshi and Mikio Inoue, who had both come from southwestern Japan to help, and several others to mark each of the 800 fish we used (Dolly Varden and rainbow trout). Each fish received a unique pattern of tiny colored lines created by injecting silicone-like fluorescent elastomer dye into certain fins. This task alone demands huge patience, and requires perfection to accurately record all the data. The rest of us went to other local streams to capture the rainbow trout we needed to supplement those removed from the sections.

The buzz and whine of cicadas was deafening as we ate our lunch in the dappled sunshine beneath the riparian forest along Horonai Stream, in a rare moment of relaxation. Tiny insects that the Japanese call *nukaka* (no-see-ums) flew and crawled everywhere, looking for any bare patch of skin to bite. Poison ivy is thick and lush in many places along Horonai Stream, and tendrils climb up tree trunks, making handholds for climbing up the steep stream banks treacherous. But the forest is also a place of uncompromising beauty by mid-June in Hokkaido, adorned in rich green as leaves of maple become fully formed. Magnolia trees are in full bloom, with huge white flowers in pink blush. Ferns have unfurled from their fiddleheads, and the groups of fronds form perfect light green conical baskets scattered across the forest floor. After four solid weeks of work for me, and six for Colden and others, the experiment was finally "in the water." On June 11, 2002, we completed preparing all the apparatus in each reach and released the last individually marked Dolly Varden and rainbow trout into their new homes. A few days later, I returned home to

Colorado, tired from the intense effort and praying that the experiment would go well and help us understand the tangled indirect effects of fish invasions and riparian-habitat destruction on these linked ecosystems.

∽

There wasn't much I could do other than hope. The e-mails coming out of Japan told of men struggling against nature, with long odds. Typhoons! Japan has two rainy seasons, one called *Tsuyu*, the monsoon season in early summer, followed later by typhoons (*taifu*), or hurricanes, that roar out of the South Pacific Ocean and move north along the islands in late summer. Our Japanese colleagues had told us that flow in Hokkaido streams is normally low and stable in June and early July, which is why we planned the experiment for that period. But 2002 turned out to be different.

There were the "normal" challenges of any field experiment. Bites from the *nukaka* left big red welts that took forever to heal, equipment broke, waders and raincoats leaked, fences had to be fixed with more sandbags, and some graduate students who promised to help couldn't come—all the usual challenges. Colden and his crew pressed on. Herons and mink decided that fish enclosed by fences must be dinner, and so apparently ate about half the fish from four of the reaches. Colden and crew traveled again to catch charr and trout to restock them. In the end, the data could not be used from the one reach (replicate) hit hardest by the predators. But then I received an e-mail that was beyond my experience as a scientist, and that forced me to add "conquering fear" to my to-do list for the research. Colden titled it "Trouble on the horizon." A typhoon was predicted to hit Hokkaido on July 6, bringing large amounts of wind and rain. Wind and rain bring leaves, and leaves clog fences, so the crew prepared for the worst.

The next ten days were hellish work, as not one but three typhoons roared north, each predicted to hit Hokkaido square on. Rain came nearly every day, in every form from drizzle to drenching downpour, saturating the soil until it could hold no more. During each typhoon, Colden and his crew worked nearly around the clock, cleaning the thirty-two fences five or more times a day to keep them from washing out. Long days of soggy work collecting samples were followed by midnight expeditions into the forest in the rain to clean fences by headlamp. As leaves clogged the fences and created small dams, the crew had to work carefully to prevent releasing large slugs of water that could scour periphyton from the bricks that we had laid out to sample for the experiment. Intermittent e-mails at odd hours chronicled the fatigue of days sampling periphyton, benthic insects, emerging insects, terrestrial insects falling into pan traps, drifting insects coming into and going out of the study reaches, fish diets, fish lengths (to determine growth), and riparian spiders (at night), all while keeping the fences clean. And after the work was finished each day, some samples needed to be analyzed in the evening, others prepared for freezing or storage, equipment cleaned and dried, and supplies prepared for the next day. Some days were eighteen hours long. In return, I could only hope, pray, and send encouraging notes of support by e-mail.

Every intense effort by a team of people needs an identity, a flag to rally around, a name or symbol that captures the essence of it. On our long drive across Hokkaido to gather the charr, we had decided that we needed an acronym to identify our experiment, and had made a game of creating it to pass the time. We would use the acronym to label equipment, make team T-shirts, and identify data sheets from our experiment. But most important, we thought, it must have a good meaning in both Japanese and English.

After much playing with transliterated forms of Japanese *kanji* that could match English acronyms, we settled on two characters that are pronounced *ishi*, which our Japanese colleagues said means "strong will" in English. We knew in advance that our experiment would require great resolve and strong will on all our parts if we were to ever complete it, so this set of characters would have a good meaning in Japanese. To match, we invented the English acronym "Invasion-Subsidy: Hokkaido Investigations," which also fit, even though it was a bit awkward. We were proud of the brand we had created for our work, which linked our two languages and cultures. However, about a week later, a Japanese PhD student happened to inform us that there are two other sets of *kanji* characters that are pronounced *ishi*. Unfortunately, these have other not-so-good meanings: "last will and testament" and "death by hanging." When the typhoons roared through in July and threatened to destroy the field research that took an entire summer of hard labor by more than a score of people to complete, we learned the true meanings of *ishi* for our experiment.

Figure 11. Three sets of Japanese *kanji* characters, each pronounced *ishi* but with very different meanings. The top set, which means "strong will," is the one we chose as a symbol with good meaning for our field experiment. However, we later found out that two other sets, also pronounced *ishi*, have meanings that are not so good.

Finally, on July 18, the e-mail came that I had been long hoping for. "The hay is in the barn!" read the subject line. They had made it through the worst, and there were only a few more emergence samples to gather and spiders to count at night for the final sampling period. Not only that, but the results of the experiment seemed clear just by looking at some of the samples, especially the samples of periphyton. Could it be that Colden and his crew had not only survived the onslaught of three typhoons, but that the experiment had actually worked? "The boys are all tired smiles, and we're all going to sleep well tonight!" he wrote. Exhausted and entirely spent, Colden arranged over the next few weeks for the samples to be shipped to our laboratory and returned with his wife and young daughter to the United States.

The entire next year in our lab was spent sorting and weighing the many samples of invertebrates and fish diet contents and analyzing the reams of data using statistics. And as the results emerged, first for periphyton and spiders, then benthic insects, emerging insects, and fish diets, they were nothing less than amazing. They fit our original hypothesis almost perfectly, something very rare in field ecology. In reaches with the greenhouses, where terrestrial insects were prevented from falling into the stream, Dolly Varden ate seven times more benthic insects than those in the control reaches and cut the mass of these insects remaining on the streambed nearly in half. The majority of the benthic insects graze the biofilm, so as a result the periphyton bloomed, increasing by nearly 50 percent. With fewer benthic insects, the mass of adult insects emerging from the stream was only about one-fifth the amount measured in the control reaches without the greenhouses, just as we had predicted. Likewise, because this emergence was also cut off from reaching the riparian zone by the greenhouses, spiders also declined to one-sixth of the number counted in the control reaches. In the field of ecology, these are all huge effects.

But the most amazing thing was that nonnative rainbow trout had nearly the same effect as creating the physical barrier between stream and forest using the greenhouse. The data clearly showed that these invaders were able to grab most of the terrestrial insects that fell into the stream and prevent Dolly Varden from eating hardly any of them. When Jess Jordan and Colden snorkeled in the reaches and observed fish feeding, they found that the trout also dominated the best foraging locations, even though the trout and charr we introduced were about the same size. These two effects caused Dolly Varden to switch to foraging more on benthic insects, just as Nakano and I had found in Poroshiri Stream a decade before when we filtered out most drifting insects using nets. As a result, the rainbow trout caused a similar reduction in benthic and emerging insects, and the same trophic cascade and increases in periphyton, as the greenhouses. In the end, simply adding rainbow trout caused spiders to decline to only one-third the number counted in the control sections where Dolly Varden were left alone. And if the invading trout could have that effect on spiders through four other links in the food chain (trout–char–benthic insects–emerging insects–spiders), we realized that they probably also could reduce birds and bats that depend on the emerging insects. As Colden made the graphs and showed me the results, I could feel Shigeru beaming and nodding in approval.

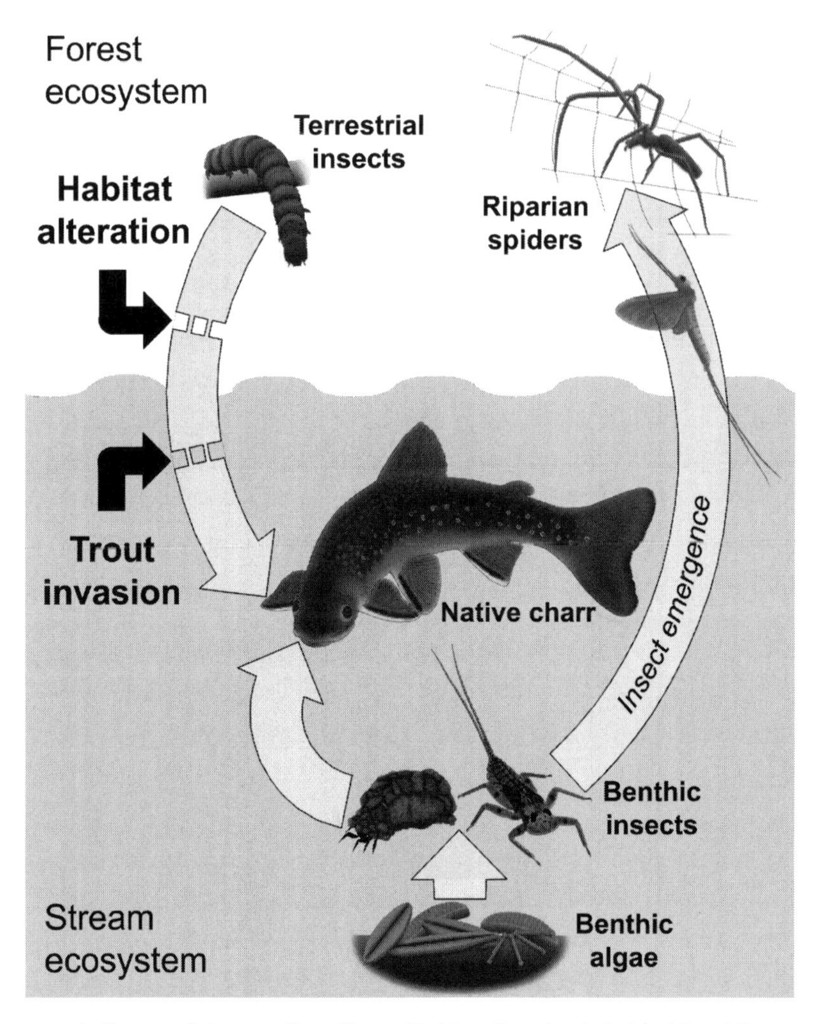

Figure 12. A diagram of the cascading effects of habitat alteration (mimicked by adding greenhouses) and trout invasion (adding rainbow trout) on linked stream-riparian ecosystems in Japan. Either factor reduces terrestrial insects available to Dolly Varden charr. For example, rainbow trout usurped terrestrial insects falling into streams, causing Dolly Varden to eat more benthic (bottom-dwelling) insects. Without these grazers, the benthic algae bloomed, and fewer insects emerged as adults, so spiders also declined. Image by Jeremy Monroe, Freshwaters Illustrated (reprinted from Baxter et al. 2004, with permission).

It is one thing to find amazing results, but yet another to publish them in a professional journal. The idea that introducing rainbow trout to a stream could potentially reduce birds in the riparian zone by subtle indirect effects on invertebrates that cascade through the food web certainly seemed unique to us. We thought it would interest many scientists. So we prepared the manuscript with great care and sent it to the best scientific journal in North America. It was returned within a few days. They were not interested. So we worked harder, rewrote and refined it more over the next month, created a new figure to clearly illustrate the indirect effects, and sent it to the best journal in Europe. It met the same fate, again within a few days. We refined it still further, and submitted it a third time to another excellent North American scientific journal. Several of the editors for this journal happened to visit the Tomakomai station when the experiment was running and were very impressed after Colden explained the design and goals. However, after receiving reviews from other scientists, this journal too rejected the work. Finally, on the fourth try after still more rewriting, it was accepted and published in the best ecological journal in North America, and Colden won a prestigious award for this work from the professional society of stream ecologists the same year. This kind of intense peer review is not uncommon in science. It is neither quick nor easy and, like much in human affairs, is sometimes not entirely fair. It is a hard road, and tests the patience and persistence of even those with the strongest will. In hindsight, however, it often results in papers becoming much better than they might have been without the refining influence. Now our project would be available for all the world to see.

Colden stopped the Toyota Land Cruiser along the dirt road where the forest of maples and scattered conifers sloped sharply down toward the spring stream. Beneath the trees was another shoulder-high forest of pencil-thick *sasa* (bamboo), much higher than the low-growing *sasa* in the mountains of Hokkaido, near Poroshiri Stream. I shouldered the backpack electroshocker and picked up the electrodes and some hand nets and started down the slope. Suddenly, I was flat on my back with a *wham!*— and staring up at the trees. The bamboo shoots were so thick that we had to push some down and step on them to make our way through, but underfoot they were as slippery as greased poles. The rest whipped at our faces and arms as we picked our way down the hill, tripping and stumbling along. Finally, we emerged near the stream, its banks lined with a lush array of tall *itadori* (Japanese knotweed) and *fuki* (pronounced *fhookey*), a huge bog rhubarb plant that is eaten as a wild vegetable. Even on this warm, muggy summer day, the cold spring stream turned the sultry humidity into a cool gentle mist, rising from the stream and softening the colors and lines all around. We had arrived at the headwaters where *oshorokoma* (Dolly Varden) still hung on in southwestern Hokkaido, against the oncoming invasion by *niji-masu* (rainbow trout).

Nearly fifty years ago, the biological mathematician Richard Levins wrote that ecologists are constantly striving to develop theories and models of how nature works that are precise, general, and realistic. Unfortunately, he noted, achieving all three of these things at the same time is a nearly impossible task. Experiments can give us precise answers about the effects of the few factors we change using our treatments, but the conditions are often not very realistic, especially if the experiment is done in the laboratory. And even those that we conduct with huge effort in the field (such as ours in 2002) can be fairly realistic, and the results can be relatively precise, but they are still conducted in only one stream. What if that stream happens to be different than most others in the region, and just a bit unusual? Then our results will not be general and won't apply very well to other streams nearby.

So, in 2003, we set out to discover whether our amazing findings from Horonai Stream were a fluke. Scientists are trained to constantly question their results. The whole endeavor rests on trying to prove themselves, and others, wrong. When a phenomenon continues to occur despite repeated challenges to refute it, then scientists believe they have discovered a bit of the truth about nature, an idea of lasting value that others can build on to discover even deeper truths. Could it be that the results we found in Horonai Stream were unique to that stream? Or, do rainbow trout really usurp terrestrial insects from Dolly Varden in other similar Hokkaido spring streams and cause cascading effects on algae and spiders? We were very curious to find out, and yet also a bit apprehensive. Many other factors can alter results out in the "real world," beyond the control of our realistic field experiment.

After searching several major watersheds in Hokkaido where our colleagues told us that rainbow trout were invading into streams with native Dolly Varden, we settled on the headwaters of the Shiribetsu River and a major tributary, the Makkari River. These streams drain the slopes of Yoteizan (Mt. Yotei), a large quiescent volcano in southwestern Hokkaido that last erupted about 3,000 years ago. Like Horonai Stream, the beautiful spring streams that drain from the base of Yoteizan in all directions have a nearly constant flow of cold water fed by the aquifer created by the volcano. They created perfect habitat for the Dolly Varden, which originally flourished in this region. But rainbow trout had been released in the rivers downstream when fishing ponds flooded, or deliberately by anglers, and had invaded upstream toward these headwaters. The nonnative trout were spreading like an ecological cancer. The streams seemed like a nearly perfect natural laboratory for testing our ideas about this rampant invasion.

But another force even more dangerous to Dolly Varden and streams than invading rainbow trout moved like a specter throughout southwestern Hokkaido, and chased us away from many reaches of stream we might have used to test our ideas. Although Colden had pored over the maps and chosen many different headwaters for reconnaissance, this force pervaded nearly everywhere we looked. It left deep tracks near bridges and roads as it galloped across the landscape and chewed along riverscapes. It changed beautiful streams from gentle meanders to straightened ditches. Shade-rich riparian forests and stream banks adorned with lush green shrubs and plants were reduced first to bare dirt and later to mown grass along boulder or cement-lined channels. The Japan Ministry of Construction was on a mission to straighten streams throughout the region.

Everywhere in the world, people want to train rivers to flow straight. The main idea is to prevent flooding of riparian areas and destruction of roads and bridges and farm fields and human property built in these floodplains. However, streams naturally meander, and floodplains are supposed to flood. Both features slow the floodwaters and absorb the energy that floods bring. When streams are straightened to drain floods quickly, it makes downstream flooding much worse. Instead of absorbing floodwaters and releasing them slowly, the many channelized headwaters in a branching stream network rapidly concentrate the full force of the flood on the single mainstem river downstream. In the United States, we witnessed this kind of destruction during 2003, when rains created floods throughout much of the headwaters that feed the Mississippi River and channelization focused the full force on the mainstem in places like St. Louis, Missouri. Since then, river managers have been working to restore floodplains that can absorb some of these floodwaters, bringing back some of the natural flood control that nature provides for free.

But even in those streams draining Yoteizan that had not been channelized, where the aquatic and riparian organisms still enjoyed relatively undisturbed habitat, we found a disturbing pattern. Where rainbow trout had invaded, Dolly Varden often occurred only in short segments of a few hundred meters of stream at the very headwaters, near where the cold springs welled up out of the volcanic deposits to form the first flow. Downstream, away from the cold headwaters, the density of rainbow trout increased rapidly. We settled on a study of six segments in three streams that originally had been inhabited by Dolly Varden only. Two segments had not yet been invaded by rainbow trout, and two located near the upstream leading edge of the invasion had rainbow trout at low density. The other two were farther downstream and had a high density of rainbow trout. Although neither control nor randomization are typically possible in such studies, replication is important, and even this small amount (two of each type) provides stronger evidence than only one of each. After all, a single segment could be a bit odd, for any one of a hundred reasons.

During this second summer field season in Hokkaido, rains intermittently drenched our work (although no typhoons came), and clouds of mosquitoes and *nukaka* tormented us at every turn, just like the summer before. And just like the 2002 field experiment, it took all the energy and stamina that our small crew had to measure the density of fish by electrofishing and to sample what the trout and charr ate, the growth of periphyton on bricks that Colden and his crew placed in the stream, and the number of spiders at the riparian margin.

In the final analysis, we found results nearly identical to those of the field experiment, with some exceptions for the algae and spiders that proved the rule. As in our field experiment, when Dolly Varden were alone in streams they captured large amounts of terrestrial insects. But when rainbow trout were present, even at low density, they usurped most of these falling insects. Where rainbow trout had invaded, the charr ate only about a fifth of the weight of terrestrial insects compared with where charr were alone. The rainbow trout apparently ate what they wanted first. In addition, at the sites with rainbow trout, the total weight of all

Dolly Varden was less than a quarter of that present where the Dolly Varden were alone. In particular, Dolly Varden fry tended to be scarce at sites with rainbow trout compared with those where the charr were alone, suggesting that rainbow trout either ate or outcompeted charr at this vulnerable early stage in their lives, or disrupted the spawning of adult charr.

Unfortunately, the results were not as clear for periphyton and spiders, the two more distant links in the food web that we measured, apparently because of other confounding factors. For example, we expected that when rainbow trout were present they would force Dolly Varden to feed more on benthic insects, thereby reducing insects emerging from the stream and the spiders that feed on them, as we found in our experiment. Spiders were indeed scarce at three of four sites with rainbow trout, compared with the two sites with only Dolly Varden, but the difference was not sufficient to provide strong scientific evidence based on statistical analysis. However, the one site with rainbow trout that had many spiders also had much more habitat that was shallow and had lots of aquatic plants. This probably prevented the charr from capturing as many aquatic insects, leading to more insect emergence and more spiders. We would have needed more replicate streams to account for such confounding factors that change the food web out in the real world.

Doing science well takes time. Although I had written the grant proposal in November 2000 to provide the time and money to do this research, the final paper from our work was not published until May 2007. Another paper synthesizing all the results appeared in 2010, a decade after the proposal. In addition, comprehensive studies like this are relatively rare in ecology. It is not common that a group of people has the wherewithal to combine the power of a controlled field experiment large enough to encompass interacting processes that drive food webs like fish foraging and insect emergence with a comparative field study that tests these findings in the messy real world. At the end, when we had succeeded not only in planning and conducting the work, but also in convincing other scientists that it was worthy of publishing for everyone to read, Colden and I both stopped for a fleeting moment to celebrate. We felt keenly the triumph of accomplishment—that we had wrested some measures of truth about the workings of nature from our relatively brief moments of time together in Hokkaido. We had answered the big questions, and fulfilled our goal to offer these findings to other scientists and managers who seek to predict what will happen when riparian forests are altered and nonnative fish invade streams. And, like Sir Isaac Newton, we knew that the main reason we had been able to accomplish these things, to see a bit further, was that we were standing on the shoulders of a giant.

∽

Why do men and women become scientists? What drives them to enter a profession where one may strive for years before results are found, and strive several more before the findings can be made known to others? What keeps them working and reworking ideas, proposals, experiments, and manuscripts only to meet criticism from their colleagues and repeated rejection and failure? How do they pick themselves up to revise a paper that has been rejected five times by other colleagues, all the time searching for that one blinding glimpse of the obvious, the logic that can show how the results add a key piece of evidence that can help resolve an important question?

In reality, some (and fortunately, these are few) become scientists because they are looking more for fame or wealth, not truths, and are more interested in the trappings of being scientists or professors. Others are clever and can conduct clever analyses, often based on the hard-won data of others. Some of these papers may trick others into believing something of value was learned, some piece of important truth discovered, though they are often akin to knickknacks and clutter wrapped up in a bit of bright paper. But sometimes I wonder whether much has been gained by this in the end, except a measure of fame or wealth.

Others follow footsteps, like me, and Colden, because it seems a natural path, where ideas are shaped by fathers and mothers, uncles and grandparents, who were seekers of truth and educators before them. My father was a professor of animal science, who struggled valiantly to complete a few research studies despite a schedule of college teaching so heavy that it would now be considered unthinkable. The administrators at his university originally forbade faculty from conducting any research, since teaching was the focus. He did so in secret, on nights and weekends. My mother was a nutritionist, who struggled equally mightily to support my father's career in science and her children's many interests, even offering ideas for my father's experiments and discussing the results.

Colden's family also included professors and teachers and farmers. In such an environment, one learns about science and the sacrifice that research requires through unspoken thoughts and feelings implied in conversations around the dinner table. Questions are asked, explanations are evaluated on their merits, and the ideas of all are honored equally, from the oldest to the youngest. Our families taught us the art of creating plausible explanations for patterns and events in the natural world (that is, hypotheses) and to relish the puzzle of evaluating them.

But the best among those who are scientists seek truth as their highest goal, beyond fortune or fame or even family tradition. Some of those who seek cleverness have argued that there are many truths, perhaps to suggest that they find no reason to seek any. Indeed, even in science, things are rarely what they seem at first. Deeper truths often lay hidden, to be found beneath the layers of the onion. But part of the nature of science is also the human sacrifice of seeking things beyond oneself, beyond one's own ego and reputation. The delayed gratification of this effort is the enduring satisfaction of seeing others succeed with whom you have shared the challenging journey, and earning the respect of those who know what it means to have struggled against those odds. And ultimately, the highest honor scientists can receive is that their ideas and the results of their experiments are eventually shown to be incomplete, or even wrong, and yet set the stage for others to ask and seek even deeper truths. The greatest success is to have provided some worn shoulders on which others can stand to take a longer view.

And for field ecologists, there is a richness to add to this experience of science. Our memories include a special camaraderie forged while camping out in forests and near streams, cooking and eating and living in close contact with the people who join in the intense fieldwork and bear the burdens to complete the science. There is no greater reward, and no more worthwhile endeavor in science, I think, than sharing the difficulties and the triumphs of research well done, centered on the flow of a stream, entrained in the stream of time.

‹ ›

"Hey Kurt, Jeremy Monroe wants to show you a video he made," said Colden, as they both walked into my office one day in mid-September

2003. "He wants to tell you about an idea he has for using it to highlight our research." Jeremy had been an undergraduate Fishery Biology major in our program at Colorado State University in the early 1990s and had recently earned a master's degree studying stream insects with another professor across campus. A quiet but curious student when he took my courses, he became intrigued by our work in Japan. An entire hour had passed one day talking about charr in Hokkaido mountain streams in an impromptu conversation following some routine questions after class. By now, Jeremy was an expert on stream invertebrates, and we were employing that expertise to help identify insects in the samples from our research in Japan.

"I think you not only have a great scientific story about how streams and forests are connected across their boundary," Jeremy said, "but you also have a great human story about crossing boundaries of language and culture and transcending human tragedy to understand these truths. I want to make a documentary film to tell this amazing story, with Nakano's life as a central theme. I think the combination will draw viewers in." I admit that I was surprised by this idea, coming from a person I had seen grow from a quiet college student to a confident young stream ecologist. But the video clip he showed us was stunning. He had captured beautiful images from beneath the water surface of fish swimming gracefully within their own element, not held in human hands after being caught and played by an angler. Even more amazing, he was able to fill the entire screen with close-up videography of net-spinning caddisflies gathering food particles from their tiny perfectly formed nets. He showed time-lapse images of stoneflies emerging from the skins in which they had spent their days underwater and pushing their wings out to prepare for their terrestrial life.

"My dream is to make a film that will draw the average person into this world of streams and the ecologists who study them," said Jeremy, looking earnestly from Colden to me. "I want to find a way to broadcast it on public television so that people can see the rich beauty and amazing diversity of organisms that are beneath the surfaces of streams in their own backyards." This idea seemed even crazier. Not only did this dream seem far beyond Jeremy's reach, but my life was completely full with teaching courses and mentoring other graduate students and a postdoc doing research, not to mention helping our own kids in high school and college. Selfishly, I wondered, "Where would I find the time to help Jeremy with a documentary film?"

But then a small distant voice called from another part of my mind, wondering, "Could such a film make a difference?" In this world where so many sounds and images demand attention, and there are so many documentary films, could this singular effort capture and hold the average person's interest for a time, bring them to understand something about the essence of streams, and perhaps help "move the needle" on stream conservation? Could it reach a larger group of people than we stream ecologists normally talk or write to? I recall saying that I was busy, but that I would help if Jeremy and Colden wanted to try.

Losing Nakano was a very personal tragedy, not only for his family, but also for his colleagues and me. So, for a moment, a bit of fear rose in my throat, because I knew that a film centered on his life must reveal some of our deepest feelings about losing a brilliant colleague and great friend. And yet, in that moment, I also realized that I was willing to take that risk if it would make a difference for conserving streams. Not long after, Colden landed a job as an assistant professor in Idaho, so Jeremy forged ahead with me as his main help. Suddenly, I had another project.

‹ ›

"Dr. Fausch, we're interested in the documentary film project you submitted, but we wonder whether the other partner agencies are also going to provide funds?" It was the program officer from the National Science Foundation on the phone. I paused, as thoughts and emotions ricocheted off the boundaries of my mind. Jeremy Monroe had written versions of the proposal to at least a dozen different agencies during the previous year to find funding for the film. I had edited them to try to help him improve them, but what did I know about making a film? Most agencies rejected the proposal, or simply didn't write back. Both of us were growing weary of the dead ends and beginning to lose hope that anyone could become excited about the idea of telling Nakano's story. However, even though no one was beating down our doors to give us money, the Fisheries Conservation Foundation had shown interest, albeit hesitant. The Japan-U.S. Friendship Commission was also still considering the proposal. We had included both of these as likely funding partners in our proposal to NSF.

So, despite my doubts, I embellished the facts a bit and said, "Yes, I think we have pretty strong support from the two other agencies for the proposal." It seemed to me like everyone was standing around, waiting for the other guys to throw their hat in the ring, and someone needed a nudge. What did we have to lose? "OK then, NSF will be able to provide the funding," said the program officer. That afternoon, the Fisheries Conservation Foundation also agreed that if NSF would fund the work, so could they. And, the third agency was soon happy to join as a partner, willing to "leverage their funds" to complete the project. Finally . . . finally, we were on our way.

Jeremy's vision of creating a story that could captivate the average person drove him and the small crew he hired through a whirlwind of filming that summer and fall of 2005. His idea was to capture the thoughts and ideas of Nakano's graduate students, colleagues, and mentors in Japan, and to record them speaking in their native language so they could express their deepest emotions. He also wanted to film all of these images outdoors, to help others understand the world in which field ecologists do their thinking and dreaming, along with their work. Colden and I each returned separately to the Tomakomai Experimental Forest, where we had conducted the experiment, for a week of filming, and Jeremy spent a total of six weeks ranging throughout Japan, including to Nakano's home town of Kamioka. There he and his crew recorded images of the rivers and streams that Nakano roamed in his youth, and interviews with his parents, teachers, and college friend Yukinori Tokuda, by then the regional fisheries biologist.

But Monroe's vision extended even beyond Japan, to the larger sphere of stream and river ecology that had embraced Nakano's work. He traveled to the Eel River in northern California to film and interview Mary Power, and to Arizona to capture images of John Sabo and other famous stream and riparian ecologists at their field sites. In September 2005, Jeremy filmed our Japanese and American colleagues snorkeling in Montana rivers as part of a workshop on charr ecology arranged by Colden Baxter and Jason Dunham, another colleague working on charr.

And in all the regions he visited, Monroe captured stunning underwater images of stream invertebrates, fish, and amphibians, often moving his camera across the boundary from water to air to show aquatic insects emerging into the riparian forest. He and his crew coupled this with images of terrestrial invertebrates falling into streams and being eaten by fish, and of spiders, lizards, birds, and bats foraging on the insects emerging from streams. These engaging scenes were interspersed with animated diagrams to explain the ecological relationships. As it took shape through the months of editing, the film began to show clearly the rich web of linkages not only between forests and streams, but also among the ecologists who study them. It seemed obvious to Jeremy that the title of the film had to be *RiverWebs*.

‹ ›

"So Jeremy, what have you learned about broadcasts of the *RiverWebs* film from that tracking service?" I wrote in an e-mail in early February 2009. It had taken several years for him to finish the final editing, and more before the film was available for broadcast by television stations. Jeremy and I had shown a "director's cut" at several meetings of our professional societies and at universities, and each time our colleagues gasped in polite amazement, especially at the underwater video sequences. After the final version was complete in October 2007, he submitted it to various film festivals and contacted our regional public television station. Our dream had been for the film to be shown by a Public Broadcasting Service (PBS) station even once. If we reached that goal, we would be ecstatic, we thought. But Jeremy's vision extended even further.

By summer 2008 the *RiverWebs* film was winning awards, and was soon shown at the largest environmental film festival in the United States and the largest in Canada. In June it was broadcast on Rocky Mountain PBS throughout Colorado. And after doggedly persistent effort by Jeremy, it was back-transformed to high definition and made available for all PBS affiliates through the National Educational Telecommunications Association (NETA) in December 2008. We had hit the big time, and Jeremy's dream of bringing Nakano's story to a broad audience had come true at last.

"Oh, yes, the tracking service shows that *RiverWebs* has now been beamed to more than 43 million homes on PBS stations in 31 states," wrote Jeremy in response to my e-mail. My mind struggled to fathom this number at first. Television stations in big markets like Los Angeles, Philadelphia, and San Francisco had chosen the film from all the documentaries and programs provided by NETA, often broadcasting it during weeks when they showed other environmental programming. Suddenly, after the years of keeping faith, we were reaching millions of homes with a unique story about how streams work, how streams and forests interact, and the blend of friendship and passion for science that leads to these discoveries.

By the summer of 2009, we had reached 70 million homes, and soon over 100 million. What if even 1 percent of people turned their television on and watched it, I thought? That would be a million people, more than I could ever reach in my life as a scientist.

∽

What had I learned from this decade of experience, working in Japan with Colden and helping Jeremy make a documentary film through which was woven not only Nakano's amazing life but also fifteen years of my own career? What had I recognized, and come to know freshly, to borrow an idea from the Western writer William Kittredge, who wrote stories about what he learned from growing up in the hard ranching life on the stark arid landscape of southeastern Oregon. What new insights had I gained after this intense and grueling work, about science, about reaching other people with that science, and about my colleagues and myself?

Although most people think of science as a cold and objective way of discovering new and useful information, in our case, and in many cases, it is much more human, and much more than cold objectivity. For us, science was also the catalyst that drew Nakano's colleagues and me back together and helped heal our hearts and souls. It offered all of us a way to make some sense of Nakano's death, endowed his life with greater meaning, and provided solace for our grief. Many of his colleagues felt a sense of peace after telling their story during the interviews filmed for *RiverWebs*, five years after his death. Many still grieve today.

I was never more proud of our work than on the day that I was able to sit in the living room of Nakano's parents beautiful home in Kamioka and witness them watch the *RiverWebs* film while Yoshi Taniguchi translated the English statements for them. I had come back to Japan to receive an international prize in my field at the World Fisheries Congress, in Yokohama in October 2008, and had resolved to travel the long distance to

Kamioka to visit Hiroshi and Hiroe Nakano. I hadn't seen them since our wonderful experience staying in their home in 1994. I wanted to show them the medal, and offer my imperfect words to convey what the award really meant for me. "This is for your son," I said, with Yoshi translating. "This is for Shigeru. Without all that he did, and all he taught me, we could never have done the research that became worthy of such an honor." Despite a complete language barrier, the pride they felt for their son after seeing the film needed no translation, and the warm thanks they offered me afterward far exceeded the modest Japanese decorum of their generation. I knew then that all the risks I and we had taken had been worth it.

I had also come to realize the quiet genius of Jeremy Monroe. People—average people—are amazed at what lies beneath the surfaces of streams when presented to them enfolded within an engaging story. They stand in awe of the intricate beauty and detailed processes revealed when a skilled underwater videographer and stream ecologist crosses the reflective boundary and records the ecological drama. Even more brilliant was Jeremy's idea to show this world through the eyes and work of the people who actually study these organisms and the ecological webs in which they are embedded, and to reveal the friendships formed and the comedy and tragedy played out in each of their lives.

In a way, people learn best when information enters through their peripheral vision, out of the corner of their eye, not viewed directly head-on but through stories. In realizing this, I also realized that the ecological connections that Nakano and I had studied were really metaphors for our own lives. We had each crossed boundaries set by our different languages, countries, cultures, and styles of doing research and found ways to have strong influences on each other, our colleagues, and our science, from an ocean apart.

As Jeremy's incredible work in *RiverWebs* began to be recognized and the film reached millions of homes across the United States, I began to think about my whole career, and what I had seen. I had gone to Japan and worked with many colleagues to study at a fairly large scale the intricate details of how several common human actions affect linked stream-riparian ecosystems. We had revealed many truths about how nonnative trout and forest habitat destruction reduce the flow of tiny insects between stream and forest, and how this change could alter entire ecosystems and cause half or more of the animals to disappear. And yet, the even more pervasive force of stream channelization was not subtle or hidden, and was marching down segments of stream and leaving them altered forever.

Pondering these things, I also began to think about my entire range of experiences closer to home, throughout the western United States, where other human actions we had studied were playing out, in subtle or drastic ways, to have unanticipated effects on streams, fish, and these linkages with the riparian zone. Were all these streams trending down, in the same way as in Japan? Were some at least stable, and being conserved?

Chapter 6
Running Dry

People live on water as much as on land.

—Wallace Stegner, *The Sound of Mountain Water*

The American West, beyond the 100th meridian, is a land defined by two things—historically, its arid climate, and recently, our overuse of its water, soil, grasslands, forests, and minerals. For Wallace Stegner, the region's native son, these were the first principles of the landscape known simply as "the West," principles that he explored through scores of essays and novels. Sadly, these two facts do not bode well for streams, the silver threads created where the water gathering beneath the land joins the soil, grasslands, and forests above. And all the while we ecologists were trying to figure out how the organisms in these beautiful silver streams of water are organized, and why the fish in them do what they do, the human race had set a course to use up the water and the land in the West at a rapidly accelerating pace. As Stegner noted, we brought our humid-land habits and technology into a dry land that cannot sustain them. And when the changes in streams in the drier parts of the region became too great to ignore, I began pondering two questions at the human-environment interface that Nakano and I had discussed the last time we met. Each passing year, these two questions grew more important. First, what is the ecological future for the streams in this dry land, and for their fish? And second, could the technology that helped us reach this pinnacle of exploitation, this engineering genius, also help us understand how we are affecting these streams and fish, and predict their future? Just at the dawn of the new millennium, we began welding these two disciplines together, engineering and ecology, to seek these answers in a very unlikely place, for a very unlikely group of fish.

∽

I recall a bluebird day on the Great Plains in eastern Colorado in mid-September 1999, with air as warm and crisp as toast and an azure sky where the only clouds were a few wispy cirrus, high in the troposphere. The waxy golden cottonwood leaves in the gallery forest alongside the Arikaree River clattered gently in the wind. The sight of the trees, in such a stark contrast of hues against the blue sky, was so vivid that I can see it clearly in my mind's eye more than a decade later. Flocks of horned larks and lark buntings wheeled low across the dry, windy shortgrass prairie beyond. Insects buzzed incessantly in the warm billows of grasses and willows along the river. It was a day to be alive!

"The pools I found last weekend are about a mile or two upstream from here," explained Julie Scheurer. I had come to help her survey the river and plan her master's research project, aimed at discovering what a rare small fish, the brassy minnow, might need to survive in a sandy-bottomed stream traversing this semiarid landscape. After sampling all summer throughout the western edge of the brassy minnow's native range in northeastern Colorado, Julie discovered that the Arikaree, which runs east to the border with Kansas, was one of the few rivers in Colorado where brassy minnow were abundant. This fish was once widespread in

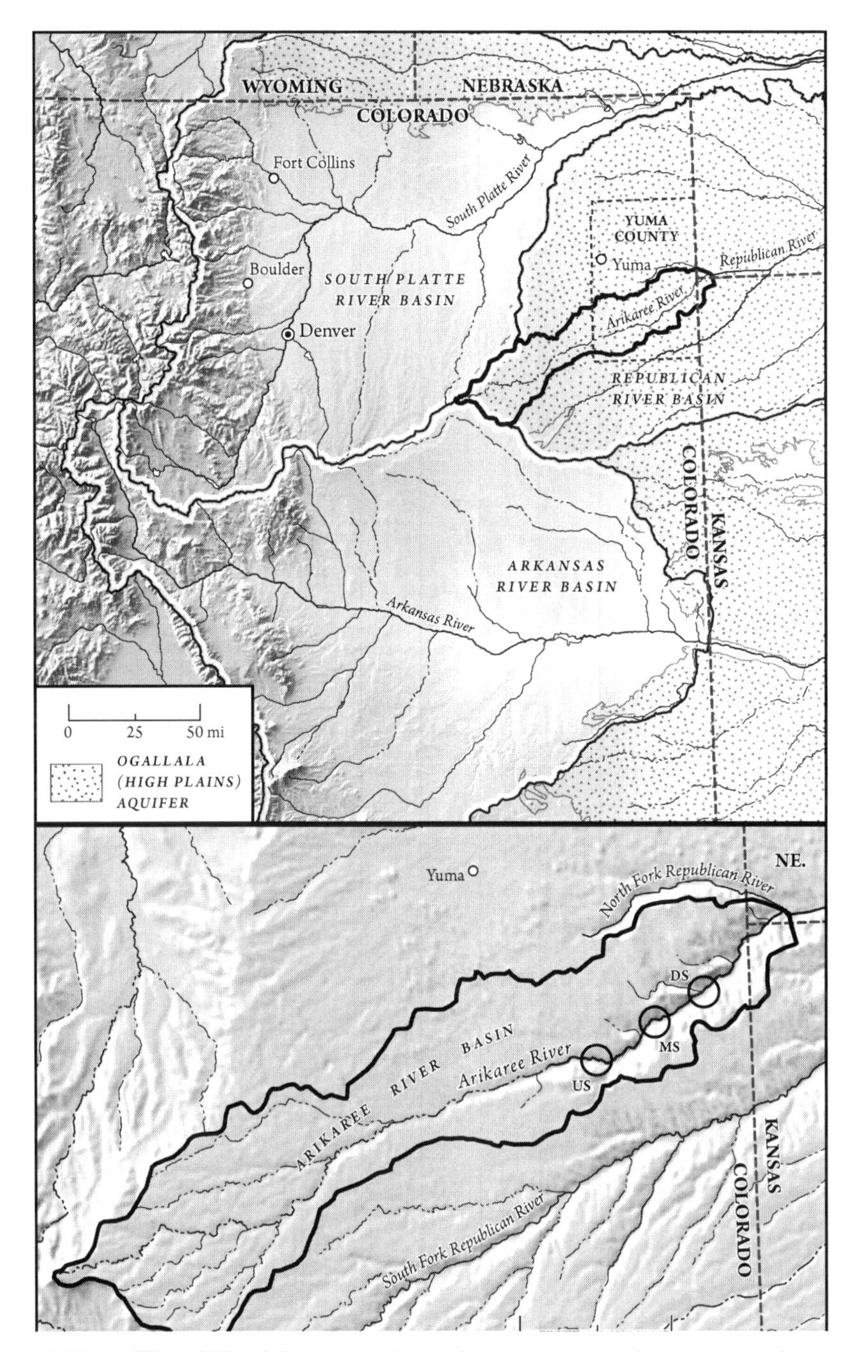

Figure 13. Watersheds of the western Great Plains in eastern Colorado. The Arikaree River basin, highlighted, is a tributary of the Republican River. The extent of the Ogallala Aquifer is shown with stippling, and the three 4-mile study segments are shown as circles in the inset (US, upstream; MS, middle segment, DS, downstream).

the South Platte River basin that covers the northeast quarter of Colorado just to the north of the Arikaree, and inhabited the tributaries as they leave the foothills of the Rocky Mountains and run through Denver, Boulder, and Fort Collins. Now, you find it only rarely, except in the Arikaree and a few neighboring streams.

Who cares about the brassy minnow? Really, very few people do care. But when a fish becomes rare, then state fisheries agencies generally add it to their list of threatened and endangered species (the "state list"). Their fervent hope is to prevent it from becoming so rare that it must be added to the federal list of species designated under the U.S. Endangered Species Act. Based on their experience with other federally listed species, state biologists know that working to conserve the populations that still remain is far easier than all the bureaucratic headaches of dealing with a "listed" species. And, to conserve populations, you need to know what the fish need to survive and persist.

Studying fish in streams on the western edge of the Great Plains can draw some puzzled looks. Many of the streams you cross on the highway look dry most of the summer. So, naturally, fish are not the first creatures that come to mind for most visitors, or even most residents. As Julie and I sat on the tailgate of our perpetually dusty secondhand state pickup and pulled on our chest waders, I mused about what the average passerby might think if they saw us. We shouldered our nets and picked up the buckets and other gear, clambered down the bank by the bridge, and started hiking up the dry stream channel for several miles. "What the *hell* do you think they're doing?" I imagined one rancher asking another, emphasizing the epithet with a Western twang. It gets pretty hot hiking up a dry channel on a warm day, even in "breathable" waders, and we joked about the paradox of this scene as we walked.

And then, after crunching along only about 100 yards through the dry crusty sand, what occurred was one of the most astounding experiences I've had in over thirty-five years of working in streams. On that day, at that particular moment, and in that place, the dry stream started flowing. Not just a trickle, but a small wave like the ones that surprise you and get your shoes wet when walking along the seashore. It quickly created a flowing sheet 3 inches deep covering the entire 20-foot-wide channel. Had we arrived fifteen minutes later, we would have found a flowing stream that left no trace of the missed event. There were no thunderstorms in the distance to explain it and only the few wispy clouds in the sky. At least now we were wearing waders for a reason. Off went the Arikaree River, pulled by gravity downstream and beyond, seeking new places to bring life-giving water. But, what was *that—in* the water?

"Look, Julie, there went a fish!" I cried, as a small fish darted quickly downstream past us in this "instant habitat." As we continued upstream, we saw more small fish, and big bulbous khaki-colored tadpoles of the nonnative bullfrog, all surfing this new wave toward parts unknown. "I wonder what my pools will look like now?" mused Julie as we trudged on, shifting the seine nets and buckets as our arms grew tired. Apparently, the Greek philosopher Heraclitus was right. You can never step into the same river twice.

Julie had purposely chosen this 4-mile segment because by late summer it was almost completely dry. Another paradox. Why look for fish where the stream is mostly dry? We reasoned that if we studied where and how brassy minnow survive and reproduce in three 4-mile segments that spanned a gradient from pretty wet (most pools are connected by flow) to mainly dry (few pools) during summer, then we could discover what threshold of habitat is needed to sustain this rare species. In short, we asked, how dry is too dry for brassy minnow? Surely, this segment would be a benchmark that was too dry to support this plains fish.

But walking farther upstream, we found another surprise. "Look, Julie, there's a pool along the far bank," I called, as I splashed that direction. "But, this was all dry last week," she said. "Yes, but there are fish in it now!" I exclaimed as we approached and could see the small fish schooling in the clear water. Our seining netted yearling brassy minnow along with other species of fishes, which apparently had all surfed there only a half hour before. Some local ranchers asked us if fish burrow down into the sand, or fall from the sky with rain. The truth is that these fish are simply prepared to go with the flow.

That experience gave me much to ponder on the three-hour drive west toward the setting sun that evening, across the rolling shortgrass prairie back to Fort Collins where I live. Why did the river begin to flow just today? Were these dry channels important to the fish? Most biologists surveying fish in these plains streams in those days would have simply recorded the bridge location using their GPS unit and marked the site as "dry" in the field notes, signifying that there was no fish habitat. "How do brassy minnow and the other native fishes persist in such a harsh environment?" I wondered. How could we ever study fish in such a dynamic stream, where segments change so quickly from dry to wet and back again, across such large expanses of space? And, as always, self-doubt crept in. Was I giving Julie good advice for her research? And could the answers we hoped to find really help biologists bring these fish back?

My colleague Kevin Bestgen at Colorado State University, an expert on the biology of plains fishes, reminds me that to understand native fish you must understand the ecosystem in which they evolved. The Arikaree River traverses lands originally grazed by huge herds of American bison, following the nutritious short grasses greening from south to north each summer and attended by packs of wolves and Native American hunters. John C. Frémont explored the region in 1842 and 1843, starting each spring from Missouri and riding along the South Platte River just to the north, recording the treeless expanses of prairie and the wide, shallow, sandy rivers.

This is the land that was homesteaded by pioneers, starting in the 1860s after the 1858 gold rush in the mountains west of Denver. More flocked there after a decade of wet years in the 1880s, displacing the bison and Native Americans. These homesteaders were duped by the slogan that "rain follows the plow," only to be driven out by multiyear droughts that caused crop failures. The cycles of wet years and immigration were repeated, eventually resulting in the overcropping and overgrazing that helped produce the raging Dust Bowl during the deep drought of the 1930s. Clouds of choking soil particles nearly 15,000 feet tall reached east to Chicago and Washington, DC. And, this is the land at the western edge of the Ogallala Aquifer, where farmers drilled deep wells starting in the early 1960s to irrigate crops, using the long sprinkler arms to create the huge green circles in the dry land. The new technology avoided the droughts and helped make Yuma County, which includes about half of the Arikaree River basin, one of the top corn-producing counties in the United States.

As for the plains fishes themselves, their family histories are set in the erosion of the Rocky Mountains to the west and the ebb and flow of glaciers to the north and east. The sand and gravel that water and ice carved from the rapidly uplifting mountains 2–6 million years ago, during the late Miocene and early Pliocene epochs, was spread eastward by wind and rivers in layers up to several hundred feet thick. Rain and snow saturated these layers of sediment, creating the underground Ogallala Aquifer that feeds the Arikaree River.

Ancient rivers draining north to the Arctic and east and south to the Mississippi River and Gulf of Mexico during this period allowed fishes from other regions to enter the western Great Plains. Glaciation during the last 2.5 million years then created cooler and wetter climates that allowed them to spread. Glaciers also diverted the courses of all major Midwest rivers toward the south, bringing their fishes to plains rivers. As a result, many species native to the region originated in the drainage basins to the north and east and are still widespread there. For example, the brassy minnow ranges northwest to British Columbia and east to New York.

The brassy minnow was also found farther south during glaciation, based on their fossil remains in southwestern Kansas, along with many other fishes from cooler northern regions. But during the last 10,000 years, after the most recent glaciers retreated, the region began to dry and warm, and only the hardiest species survived in plains rivers. In fact, all of these hardy plains fishes can withstand water temperatures up to at least 93°F and low dissolved oxygen levels down to 0.5 parts per million, enough to kill or seriously stress most other fishes from less arid regions.

It is not surprising that we know little about these western Great Plains fishes in Colorado. The first fish preserved for study were collected during railroad surveys in 1873, and only twelve sites were sampled before 1900, all near the foothills of the Rocky Mountains to the west. However, farmers began diverting water for agriculture in earnest in the 1860s in the South Platte and Arkansas river basins, to the north and south of the Arikaree, and this early habitat loss probably drove some species extinct without a trace. In fact, several fish species were caught only once before they apparently disappeared completely from the region, never to be seen again since. In the Arikaree, only a few sites were sampled by 1940, and the first widespread sampling was not until 1979, so some species also may have been lost from this basin before any specimens were collected.

We still don't know the most basic biology of some of the fishes that inhabit these plains streams, such as where they spawn, what they eat, and how old they live. And, although these fish are incredibly tolerant, they are steadily declining. Of the thirty-six native fish species ever captured east of the Continental Divide in Colorado, nineteen are either extinct or on the state list of species that are endangered, threatened, or of special concern (the brassy minnow is a threatened species in Colorado). Of the sixteen fishes known to be native to the Arikaree River, only nine remain. Seven have disappeared as of this writing, two during the eight years that we worked there. More than half of the fishes in the region are gone or in trouble, most of them deep trouble.

How could fish this hardy disappear, we wondered? More importantly, how could we understand what habitats they need to persist into the future, especially across the large spatial scales we apparently needed to sample? And, how could I help my students study this problem in this riverscape without driving them to insanity and exhaustion?

As a field ecologist, sometimes you have to take a gamble and use your intuition when planning research, rather than relying on approaches used by other scientists or choosing the safe route that would give the right answer to the wrong question. To those who are not ecologists, this gamble may seem unscientific, but then we're not the kind of scientists who wear white lab coats or conduct neatly controlled experiments in cloistered calm.

Previous biologists had typically measured stream fish in great detail in 50- to 200-yard reaches, often through different seasons and over many years. But my intuition for Julie's work on brassy minnow was that we needed to think big. We needed to study them in less detail but over much longer expanses of stream, segments as long as we could manage. This hunch came from my work with earlier graduate students in the 1980s and 1990s and the research of several other fish biologists. Those studies revealed that even stream fish we assumed to be "residents" moved long distances, sometimes miles, even in small streams, to find the habitats they needed to complete their life cycle. We learned that the places adult fish spawn, and habitats where juveniles feed and grow, are often distinct and far apart along the river network. These habitats are also often far away from the places that fish need to survive over winter. Many animals, if not most, from our Paleolithic human forebears and Serengeti wildebeest to plains fishes, need (or needed) to move to find the right places to sustain life at different stages.

An earlier study in a stream a few hours to the south of the Arikaree showed us why moving is so important. During his master's research, Ted Labbe discovered that the Arkansas darter, another small threatened plains fish, moved up to three-quarters of a mile seasonally between pools to find the best places to survive and reproduce. Some pools that were warm and productive allowed the earliest spawning and highest growth of tiny newborn darters during summer, but then froze to the bottom during winter, killing all the fish. In contrast, another pool where adults survived the winter was completely ice-free because it was fed by groundwater springs, but then was too cool in the summer to allow the best survival and growth of darter larvae. Fish that spawned in the warm pools in summer, but whose offspring moved to the cooler pool for the winter, could find the best of both worlds.

Ted also studied even longer expanses, up to 21 miles, and found that long reaches dried up during summer, leaving hardy plains fishes crowded into the shrinking pools that remained. Whether fish become meals for raccoons and birds, die when the pool dries up, or are saved by thunderstorms that restore flow just in time depends on the lottery of towering cumulus that parade across the plains nearly every summer afternoon and drop rain in random locations. The strange thing is that fish apparently don't have the sense to seek out deeper pools that provide refuges during this summer drying. However, Ted found that the 2-inch-long Arkansas darter could quickly recolonize these long reaches when water returned, apparently moving up to 2 miles from the nearest location that sustained populations.

"I think we should sample 4-mile segments, which should be long enough to encompass the life cycle of a 3-inch-long brassy minnow," I said to Julie, knowing full well that "we" meant primarily her. Sampling three of these segments that ranged from wet to dry (arranged upstream, middle, and downstream along the basin) would be a big challenge—a lot of grueling work. A bit to my surprise, Julie embraced the plan, swallowing any of her own doubts. But then, she had done challenging fieldwork before, in big brawling Oregon rivers. Sizing up what little we knew, and using my intuition to go big, we started by asking the simplest questions we could imagine. For example, during each season, what reaches of the Arikaree River were wet, intermittent with isolated pools, and dry, and when did they become connected and disconnected? Where did adult brassy minnow live and thrive, and where did they die out? When and where did they spawn, where did the larvae grow, and when did they leave these habitats as juveniles? How fast did fish recolonize dry reaches that became wet, and where did they come from? We surmised that if we knew answers to questions like these, we could figure out what brassy minnow needed to persist at the larger spatial scale that we had come to call the "riverscape."

The sun beat down mercilessly on an August 2000 morning as Julie and I started upstream along the middle segment (the one with intermediate drying) to measure habitat and sample fish during the driest part of the summer. Her technician dropped us off at the downstream end and drove the truck around to the halfway point 2 miles up. He would meet us by walking downstream. Reports on the radio warned people to stay hydrated as the temperatures soared to over 100°F in eastern Colorado. Humans actively working outside at these temperatures need to drink at least a quart of water an hour, they said. We had decided that the best way to map the habitat and find out where brassy minnow occurred in each season was to walk each segment, record what was wet, and sample every pool for fish, or at least a large random subset of them. Julie did this several times each year in the three 4-mile segments, for two years. Earlier work she conducted showed that we could detect whether brassy minnows were present in a pool by making two passes through it with a seine. Even if only two brassy minnows were present, we calculated that we were almost certain to catch at least one during the two passes (98 percent of the time). Therefore, if none were caught, we could conclude that they were most likely absent from that pool.

We worked upstream steadily through the treeless vista of rolling shortgrass prairie, pool by pool. After seining each pool, we sorted brassy minnows from the hundreds of other fishes, counted and measured each one, and recorded the data on forms in our metal clipboard. The sun grew high and the heat was intense, reaching over 105°F. Rivulets of sweat carried sunscreen into our eyes, and wiping it away made them sting more. Would the two quarts of water we each carried sustain us until we reached the truck for lunch? Or would we wither and faint, and meet the same fate as the fish in drying pools? Fortunately, the pools grew sparser along this lower part of the segment, and the work went a bit faster. At lunch, a dry prairie breeze cooled us slightly, evaporating the sweat from our soaked field clothes as we rested in the shade of a few sparse cottonwoods near the truck. However, that afternoon, in the still air and baking sun, fatigue set in from the long day of extreme heat, and we started stumbling and working more slowly. My mind wouldn't think clearly when recording the data, even after drinking plenty of water. Julie looked flushed in the heat. We barely completed the last pools before we felt we could sample no more, and wearily trudged the 2 miles back to the truck.

In the end, the fish proved tougher than we were. Over the two years of her research, Julie found them in pools where the water reached 96°F in late afternoon. The oxygen dissolved in the water dipped to 0.03 parts per million when she measured it at dawn, after the respiring plants and bacteria had sucked nearly all the oxygen from the water during the night. Most fish in wetter regions need 5 parts per million of oxygen or more to survive. Several years later, as PhD student Jeff Falke began his dissertation on the river, we learned that eastern Colorado began sliding into a long drought in 2000, reaching the worst in 2002 and not recovering until after 2007 when Jeff finished. The Palmer Drought Severity Index used by meteorologists showed that this drought was nearly as severe as the Dust Bowl, and nearly as long. Though very stressful for the river and its fishes, this long drought allowed us to learn much about how both respond to such dry conditions.

By fall 2000, after Julie had sampled the three 4-mile segments for three seasons, we realized we wanted an even larger view of the river. I knew I had tested the limits of her stamina and commitment, but I still wondered what was happening *between* the segments. "It sure would be great if you could fly and look at the whole river," I remarked one day. Christian Torgersen and his colleagues had just published research in which they had used helicopters to videotape Chinook salmon habitat along an entire 75-mile river in the dry country of eastern Oregon, using infrared videography to measure stream temperatures from the air. But we just wanted something "low-tech," to see what was wet, dry, and intermittent, and to help place Julie's ground surveys in the context of the whole riverscape. How much habitat did the fish have, really, throughout the whole river, especially during the driest part of the summer?

"The guys at the Colorado Division of Wildlife said their pilots that fly big game surveys and stock fish in high mountain lakes are sometimes not that busy," said Julie cheerily, rising again to my challenge. Our contact, Tom Nesler, agreed to the flights and somehow found the funding to pay for them. I can imagine the questions he must have endured when he told his superiors that we needed to borrow a plane to study stream fish. But Julie's three flights in 2000 and 2001 were so informative that they became a regular feature of Jeff Falke's research during 2005 through 2007. By 2007, he was making flights every month from May to October.

∽

"Look at that!" I exclaimed in shock, as Jeff clicked his mouse through the digital maps of the river every month from his 2007 flights. During Julie's work, it had been surprising enough that the downstream segment dried out so quickly each spring based on her ground surveys. Even though the entire segment flowed in April, three-quarters of the habitat had dried by June, and only a few pools were left in the whole 4 miles by August at low water. Jeff continued these foot surveys for three years, and the whole set showed clearly how dynamic the Arikaree was across this gradient of drying—not only during each summer, but also from wet to dry years. For example, although a quarter of the middle segment was dry in August 2005, the next year was even drier, so that nearly three-quarters was dry in August 2006. Even in the upstream segment, the wettest of the three, a third of the habitat was completely dry in August 2006.

But the flights were even more revealing. Jeff found that only one flowing segment remained along the whole river at low water during late summer in 2006 and 2007, and it was centered on our upstream segment. Starting from more than 39 miles of connected habitat in April 2007, each click through Jeff's maps made the blue-colored "wet" habitat shrink and the red-colored "dry" reaches grow. Click, click—going . . . going . . . and, in the end, most of the river was gone. By July of both years, the river had dried to a relatively short segment of core habitat in the middle, and remained that way through October. Of the entire river that once flowed about 70 miles to its mouth in the 1940s and supported fishes throughout, only a disconnected fragment about 7 to 9 miles long remained flowing during the low water period of late summer in 2006 and 2007, a mere 10 to 15 percent of the original. Otherwise, there were only a few scattered remnants of pools and shallow runs throughout the rest of the river's course. Could a fish species like brassy minnow persist under such conditions? Was this drying caused by the drought, or some other culprit? We realized that these were critical questions, but in the meantime we needed to work out the biology of how fish lived in this harsh environment.

"Time to go . . . ," called Alex Klug, rustling around in the ranch house kitchen to get his food and water ready for the night of sampling ahead. I felt like I had just lain down, after a shower and dinner that capped a long day of hiking in the heat to survey the middle segment in July 2007. Jeff Falke was taking a much-needed short vacation, and I was the temporary worker helping his technician collect the data that week. I've never really liked sampling streams in the dark, but sometimes you have to steel yourself against the elements to discover essential things about fishes. Many things worth knowing have their price.

We wanted to find out whether these plains fishes spawned in the segments, and if so, where and when. This would help us understand when and where water was most critical to sustaining populations of these fish species. So for his dissertation research Jeff set out to sample their larvae frequently during April through July, the spawning and rearing period, over three years. He used two methods, dip-netting during the day in the flooded grasses of backwaters and channel margins where we had found adults spawning in March, and setting larval fish traps lighted with glowsticks in the backwaters at night. Amazingly, tiny fish larvae less than a quarter inch long are attracted to these light traps from more than two yards away, and can reach them quickly.

Hiking across these long segments at night forces you to use all your senses, and all your knowledge from hikes during the day, to avoid becoming lost or injured. After all, a good headlamp only allows viewing the small world within the limits of its beam. You must compare this view against your visual memory of daytime scenes and the internal map you carry in your head of the whole segment. I feel rather like a beetle must, wandering through scattered grass across the plains, casting this way and that to find a path forward. "Didn't we turn here by that big cottonwood to find the fourth backwater?" I call to Alex. And just where was that strand of old barbed-wire fencing that could rip our waders? Then, oops, I forgot about that root that we tripped on during the day, so now I'm falling face first. Instinctively, I roll toward one shoulder to break the fall, and fortunately thump into relatively soft grass. I quickly pick up the buckets and nets and run a bit in my bulky waders to catch up with Alex's small hemisphere of light disappearing into the darkness ahead.

When we arrive at each backwater, we stake out one of the floating traps, or two in the large backwaters. After all the traps are laid out, we return to the truck to allow them time to attract the tiny fish. Two hours is long enough for the traps to become crammed with fish larvae and aquatic insects, but not so long that the bigger fish and insects begin eating the little ones. So, after a brief snack and conversation with Alex about his plans for grad school, we're hiking the miles again to retrieve the traps, preserve the specimens in alcohol, label the vials, and record all the data. Dead tired, we make the long hike back to the truck with all the gear, stopping once in the early hours after midnight to stare in wonder at the vastness of the swath overhead awash with the Milky Way. Amazing views of the stars in the night sky are one of the unexpected joys of working through the night in such remote places, miles from any lights.

At the end of my time helping Alex, other than ten scattered hours reserved for eating and sleeping, we were generally hiking and sampling the other fourteen hours, day and night, for four days. I concluded that this regimen constituted the perfect "Arikaree diet," for anyone like myself who wanted to lose a little weight that summer. Better yet, this sampling trip helped me appreciate even more than I had the grueling work that

Julie and Jeff and all their technicians completed without complaint to gain hard-won data on these underappreciated plains fishes.

So, what did we find out about the Arikaree fishes from all that hiking and sampling? After more months of laboratory work and analysis, which usually takes about three times longer than the field sampling, Julie and Jeff each settled down to distill our knowledge into manuscripts so we could tell other scientists worldwide about what we had found. The species of fish that were still present during Jeff's work began spawning at the end of March when water in the flooded channel margins and backwaters had warmed to the right temperature. Fish larvae lay down a thin layer of bone every day on tiny bones the size of sand grains in their inner ear, called "otoliths". When these bones are sliced and polished, you can count these daily rings under a microscope. Jeff and his technicians collected about 11,500 larvae of all fish species during his three years of research and aged over 500 brassy minnow larvae using otoliths. This allowed him to determine the dates when these larvae had hatched, and by backtracking from these, the dates when the adults must have spawned. He also knew the ages of the oldest larvae captured in the backwaters, which told us when they must have moved from this spawning habitat to begin their lives as juveniles in the main stream.

Although the locations of some backwaters suitable for spawning and rearing shifted from year to year, appearing at random along the segments, adults and larval fishes colonized them very quickly, even from long distances. Remarkably, Julie had found brassy minnow larvae throughout the downstream (driest) segment both spring seasons she worked, even though most of it dried up and killed all fishes by late summer. Likewise, Jeff found larvae in this segment that were older than the backwaters where they were captured. They had apparently drifted downstream in the flow produced by a thunderstorm, and colonized the backwaters just after they formed. We began to realize that even 4 miles wasn't long enough to encompass all the habitats these fish might use during their life cycle in the intermittent and dry segments of the river. We went big, but it wasn't big enough.

Relatively few fish larvae of any species survive their first few weeks of life, but the odds are even worse in plains streams. The brassy minnow larvae that do survive move from rearing habitat in backwaters to the main channel when they have grown to three-quarters of an inch long (about 20 millimeters), and then either find a suitable refuge pool or risk death from drying in shallower habitats. Adults live only about three years but disperse rapidly and can produce large numbers of eggs and larvae when water conditions are favorable. In the end, their game plan for survival of their species is simple. When there's water, they go everywhere and have lots of kids. Then, they and the kids that survive to adolescence hunker down in refuge pools during the summer drought, hoping that their particular pool doesn't dry up, or freeze to the bottom during the winter. Like all of us, they keep running to survive, moving to new places as they become more favorable than the old ones. In the parlance of ecology this is called a "metapopulation," a set of populations in shifting patches of habitat that are connected by movement. This shell game is how brassy minnow and many other plains fish species persist in streams like the Arikaree. It's why the dry segments are also critical habitats, because these are the future corridors needed by fish to recolonize the wet patches when the water comes.

∽

"I think we're going to have to study groundwater, to understand what drives plains fishes and what we'll need to do to protect them." I pitched my new idea to Tom Nesler in 2004, by then leader of the Aquatic Native Species section for the Colorado Division of Wildlife. Tom had conducted management-oriented research himself for many years, and now championed the cause for native fish statewide. He must have sighed inside, wondering whether my idea was too academic to produce anything useful for his fisheries managers. "The fish need pools that don't dry up during the summer and also don't freeze solid in the winter, and it's groundwater that sustains those pools," I went on. "We know that from Ted Labbe's and Julie Scheurer's studies, so if we want to make progress, I need to work with a groundwater hydrologist and someone who knows how those big center-pivot sprinklers affect the water table." Tom agreed to help, and in the end the answers proved mind-boggling. I began working with university colleagues Deanna Durnford (groundwater hydrologist) and Ramchand Oad (agricultural irrigation engineer) and their students. After a graduate student working with Ram read Julie's thesis in 2001, they had begun working out the effects of agricultural pumping on the river, using the threatened brassy minnow as part of their justification. Although I knew nothing of their work at first, like many busy professors we connected through our students.

It was not news that the water table was declining in the Ogallala Aquifer, which stretches from South Dakota to Texas and covers parts of eight states. Declines of more than 300 feet had been recorded from wells in north Texas, and by the early 1980s the water table in the Arikaree basin had dropped more than 25 feet in some locations. When permits were issued to allow drilling wells throughout the Republican River basin in eastern Colorado (of which the Arikaree is one tributary) starting in the 1960s, the State Engineer's Office knew from work by early hydrologists that the aquifer was a finite resource. They planned to allow 40 percent of the underground water to be "mined" over twenty-five years, which they considered to be the useful life of the aquifer.

Pumping went well at first and created a highly productive agriculture, but by the 1970s it became clear that the water table was declining rapidly, so a moratorium was set on new wells. In the early 1990s the plan expired so the date was extended to allow pumping for 100 years to reach the 40 percent depletion. But by 1998 it was clear that flow in the rivers had also declined, and the wells were to blame for depleting the groundwater that sustains them. So Kansas sued Nebraska in federal court, and Nebraska then sued Colorado, for allowing too many wells that pump too much groundwater, thereby failing to provide sufficient river flow to these states downstream. In short, they were using more water than they were allocated under the 1942 Republican River Compact. In 2003 the U.S. Supreme Court ruled that no new wells would be drilled and ordered the states to work together and use the same computer model to allocate water and resolve the issue without suits. As of 2011, Colorado had "retired" 30,000 of 580,000 acres of irrigated land (about 5 percent) by offering payment to farmers who would volunteer to permanently retire their wells and stop pumping. However, during the period between 2003 and 2008, Colorado had still used about 40 percent more water than was allocated under the new agreement, in part because much of this period was during the long drought.

"My intuition tells me that we should just try to calculate when the last pool in the river will dry up," I said to Jeff Falke early in 2007, while staring wistfully at the many diagrams we had drawn on the whiteboard.

After all the sophisticated analysis we had already done, the stark reality was hitting me that it would all come down to this. If fish could move anywhere to spawn, and later sufficient numbers could find refuge pools, then what we really needed to find out was how fast these refuge pools disappeared, and when the last one would be gone.

We had struggled through countless meetings with our engineering colleagues and their graduate students, trying to mesh what we wanted to know with what their models could provide and to understand each other's scientific terminology and methods. They had studied the geology and water tables of the entire basin and knew how the water moved underground, seeping inexorably from the Ogallala Aquifer into the sand and gravel beneath the river, and then into the channel itself to create flow. They knew the locations and approximate pumping rates of the hundreds of wells and could simulate what effects each well had on the water table using a sophisticated computer model. But it hadn't been easy, and took four years of research by our six graduate students. In the end, it had come to this straightforward question, making the final analysis a blinding glimpse of the obvious. When working across disciplines, what fellow scientists often need from us is not our most complicated "rocket science." Sometimes, it's even a back-of-the-envelope calculation.

But to complete the computer model, Jeff needed one more set of data—the location and maximum depth of every one of the 218 refuge pools in the upstream segment, pools he had visited many times before. We focused our analysis on a 30-mile slice of the basin centered on this segment, because Jeff's flights showed that the upstream segment alone contained most of the remaining wet habitat along the entire river at low water in late summer. The year before, Jeff had proven from his field research that the water level in each pool was identical to the groundwater level near the edge of the river, by monitoring both for six pools as they dropped during summer. It was clear that the pools were sustained by groundwater. Therefore, we could run the computer model forward through time, with all the wells pumping, and simply determine what year the water table fell below the bottom of each pool in late summer.

And when graduate students Angie Squires and Robin Magelky created, tuned up, and ran their sophisticated state-of-the-art groundwater model, and Jeff analyzed the results for each pool (imagine several years of mind-numbing work here), the answer was sobering. If the wells continued pumping at current rates, and even without another drought, half the refuge pools remaining in this wettest segment along the river in 2007 would dry up by 2035, and only a third would be left by 2045, about thirty-five years in the future. Not only that, but by 2045 nearly all these pools would be concentrated in only about two-thirds of a mile of river channel, and very susceptible to drying up completely in a subsequent drought. Removing all the wells within 3 miles of the river, a solution offered by the State Engineer's Office to meet the Compact requirements, was not much better, leaving a bit more than half the pools by 2045. Water levels in the Ogallala Aquifer throughout the Arikaree basin were declining 10 inches each year, which translated to drops of 2 inches per year in the sand and gravel next to the river. Most pools were about 2 to 3 feet deep by late summer in 2007. We could all do the math. Like most science, once all the groundwork is laid and the data are collected and analyzed, the answer seems all too obvious. But building the scientific foundation to conclusively prove these conclusions can take years of struggle.

The final analysis led to broader epiphanies. At their best, good scientific theories and models can unite previously unexplained observations, and portend the future. First, the reason that the downstream segment dried out so quickly, and started to flow that September day in 1999, suddenly became crystal clear. In addition to the wells sunk deep into the regional Ogallala Aquifer, along this segment there were also shallow wells drilled into the floodplain within a quarter mile of the river. These wells intercepted the groundwater before it could seep into the river and caused the channel to dry out quickly in early summer. And when the pumps shut down at the end of the growing season in September, and the water-loving cottonwoods also began to drop their leaves, the water table rose above the bed of the river and it began to flow. Otherwise, only a large rainstorm could fill up the sand beneath the floodplain and create flow during midsummer.

Second, time is running out, and fast. Throughout the river, the habitat is drying so quickly that a new state fisheries biologist starting work in this region can't pass this problem on to the next biologist when they retire. The river will dry up and most of the fish will be lost during their thirty-year watch. And third, there was no reason to add climate variability owing to "climate change" into our model. It will only bring the end sooner than we estimated, when the next severe drought comes, probably within the next twenty-five years.

As if this weren't enough, when our group analyzed these effects more broadly, we found more bad news. "We can use a simple water-balance model to calculate how much the pumping would need to be reduced just to keep the habitat we have now," said Deanna Durnford, after pondering the problem only a minute. It was nearly one of those back-of-the-envelope calculations. Again, the answer was stark. At least three-quarters of the pumping would have to be stopped just to keep the short 7- to 9-mile segment of "core habitat" during late summer, which provided the ecological future for the nine plains fish species remaining in the Arikaree, including the brassy minnow. Even more pumps would need to be shut down to let the water table rise and create more stream habitat to buffer against future droughts. However, the value of agricultural sales in the Arikaree basin totaled about $400 million per year, so a change this large would require long-term planning to avoid major economic upheaval.

Adding to all of this, two simpler but broader analyses by Jeff and our colleagues in Kansas showed that these groundwater declines, driven by pumping about 100 billion gallons per year in the Arikaree basin alone, are widespread across the Republican River basin, and indeed the entire western Great Plains. For example, throughout the vast expanse of western Kansas fed by the Ogallala Aquifer, four of every ten stream miles dried up during the fifty years from 1955 to 2005 as the groundwater was pumped, and if this continues a total of six of ten stream miles will be lost by 2045. This is clearly not a local problem, confined to the Arikaree.

∽

We pull up to the huge crop circle in two university vans late on a Saturday afternoon in early November 2008. Just then the setting sun emerges below the thick layer of gray clouds and a shaft of bright rays turns the tops of the dry cornstalks to pure gold. "He should stop when he gets to the end of the row with his combine," says William Burnidge, a conservation biologist with The Nature Conservancy, as he steps from his truck. The big green and yellow John Deere combine rumbles to the edge of the corn, gobbling in rows of cornstalks. When it stops, the tractor following beside it heads off to empty the grain wagon, now completely full of the dry yellow and orange kernels of corn gleaned from the cobs by the big machine. After a three-hour drive from Fort Collins starting at dawn that morning, my class in Conservation of Fish in Aquatic Ecosystems has already visited The Nature Conservancy's Fox Ranch, in which our upstream segment lies. We seined native fish from the river to hold them gingerly and steal glimpses of their simple beauty, and learned about ranching cattle and water use for agriculture from William and the rancher who leases the riparian pastures to graze his cattle. My students began to appreciate how conserving these native fish requires knowing the details of groundwater laws in Colorado, and how the river and its groundwater are used to raise grass and grow corn and alfalfa that feed cattle. We stand miles from the river, looking at a harvest grown with water from deep beneath the land.

We all shiver in the west wind blowing from the Rockies across the high plains, as it rustles the dry corn stalks, heavy with ears. I had waited too long to arrange this field trip, which forced us to come in late fall. The entire year's income for a farmer hangs in the balance at harvest—the tractor payment, the money to pay the crop loans and insurance, and the future college tuition for their kids. Now they live and die by the weather. Wind will knock the ears of corn to the ground, where the combine can't pick them up. Rain will wet them, forcing farmers to buy more natural gas to dry the grain after harvest. Aaron Frank and his father and partners farmed fourteen of these half-mile-diameter crop circles that year, and they had to get the harvest in. There was no time to talk to college students in October. But the weather hadn't cooperated, so our November date arrived and they were still harvesting corn.

My father had been in agriculture, raised on a farm in Minnesota, and had become a professor of animal science, so one might say that I grew up at the edge of the field. I felt this helped me understand farmers a bit.

They live on the land, know its seasons, and mark the ebb and flow of life-giving water for their plants. They know when and where the streams and rivers flow that cross their land, and may even notice fish and when they move if they are large enough to see. They know what the past was like, from old family photographs and stories handed down, and where there is no longer water enough in the river for swimming, or for cattle. I know that farmers are independent, like my father was, and proud of their work, the way of life they cherish, and the food and fiber they grow to sustain us all. But I worried that my students wouldn't be so understanding. Several in the van I was driving had made a point about a recent documentary film that parodied our reliance on "King Corn" and the ills it has created.

Farmers have a second full-time job as economists, an obvious point that dawned on me only recently. Farmers are also "way into" technology. When I ask Aaron to explain his farming operation for the students, he quickly runs down the costs for seed, fertilizer, fuel, and electricity to pump the water, right off the top of his head. He talks about locking in the price of corn (futures) and how that can make or break a farmer any given year. Crop damage insurance, disease, and the growing demand for steel from China all come up. He explains how modern tractors can plant seeds and inject fertilizer to within an inch of each other beneath the soil with GPS-guidance systems, and how farmers can save water by tilling the soil less frequently and turning off the center-pivot sprinklers from their cell phone when it rains.

Aaron and his contemporaries majored in fields like Mechanical Engineering at my university and are as skilled at using computer spreadsheets as greasing and driving combines. And, they are proud—proud that they can work with the elements and buffer against adversity every year to grow crops in a challenging landscape while maintaining the soil that is their livelihood. They strive to put food on their own tables, but their real goal is much bigger. It's interesting how these passions lie just under the surface and at the margins of thought, like a mantra behind a veil in one's mind. For my students and me, our passion is to conserve ecosystems that support native fishes. For Aaron and his colleagues, it is to feed the world. Neither group expects to get rich.

By now the sun is gone, the dusk is dimming around us at the end of a long day, and I'm thinking that the students must be cold, bored of learning about agriculture and the economics of farming, and anxious to find some fast food (including beef or chicken grown with corn) before the long drive home. I ask if anyone has more questions. To my amazement, one of the students who was critical about corn agriculture before we arrived asks, "Can we see your combine?" Aaron responds like a young man with a newly restored '55 Chevy, opening every door and operating every lever, showing students the magic of how this huge machine turns corn stalks into grain. Afterward, as we load up the vans and drive the 45 miles to Yuma, students talk excitedly about what they learned. Even though their majors are in fish, wildlife, or conservation biology, the murmurs are not about the cool fish we saw that day, or the way the river looked, but instead about what they hadn't realized about agriculture, how it is done, and how it seems like farmers risk financial ruin every year to feed us all.

In the end, I can't help but feel like a few of us are standing in the wheelhouse, pleading with the captain, hoping to convince him to begin turning the Titanic so that it will hit the iceberg with more of a glancing blow. The future is bleak for the river and all the aquatic and terrestrial organisms sustained by it, including the native plants, amphibians, wild turkeys, songbirds, and deer that live in the band of riparian vegetation along it. At the end of the day, when we quiz Aaron and his partner about how they view the crisis, we find that they hold a more pragmatic, shorter-term view than we do. Cities want to buy their water to grow houses and lawns along the Front Range, but farmers would rather use it to grow crops. Some of their fathers already sold their farms and water close to the cities and moved here to continue farming, a difficult but rewarding lifestyle that they cherish and want to continue. The seed companies promise new varieties that will use 30 percent less water, they say. Farmers and The Nature Conservancy are working together to test new computer technology that controls sprinklers and applies water more carefully only where it is needed on fields, to conserve water while optimizing the yield of crops. Maybe we can make it last for a few more years, or decades.

The Colorado Division of Wildlife and other state and federal agencies have no jurisdiction over these private lands, unless one of the fish or wildlife becomes listed as a federally endangered species. This would create a bureaucracy that many state biologists find ponderous, and even a "listed" species would not guarantee that the river could be conserved. The Nature Conservancy can manage differently on their own Fox Ranch, where the core fish habitat remains, but the problems lie many miles beyond the river, where the pumps draw from deep within the Ogallala Aquifer. And now Colorado must deliver the prescribed acre-feet of surface flow to the Kansas state line.

The solution settled on by the State of Colorado to satisfy the Compact was to buy land with fifteen deep wells north of the North Fork Republican River, the next tributary north of the Arikaree, near the Nebraska state line. The plan calls for pumping the groundwater that was formerly used to grow crops on this land, running it through a pipeline, and discharging it into the North Fork at the state line, at a total cost of $71 million. This will allow the rest of the pumping that sustains the productive agricultural economy to continue, while still meeting the requirements of the Republican River Compact to deliver water to the downstream states. However, this short-term solution effectively bypasses the river and results in continued declines in groundwater throughout the basin, further hastening the loss of fish and other organisms.

During the decade that we studied the ecology, hydrology, and agriculture in these basins, and observed the politics, the most common response from federal agencies to which we applied for grants was that this was a local problem, not worthy of their consideration. About a third of the water used for irrigation in the entire United States comes from the Ogallala Aquifer, and a quarter of the world's grain supply comes from the Great Plains ecoregion of North America, of which the Ogallala makes up a substantial part. When we submitted this work for publication, reviewers for a major applied ecology journal replied that our research was not relevant, and not worth publishing, because the result was so obvious. Few, if any, studies we know of combine these three scientific disciplines over such a broad set of spatial scales to project the effects of groundwater pumping for agriculture on rivers and their biota. In the end, none of the pieces in this grand puzzle seemed to match.

So, Jeff and I and our colleagues pondered the options. What could be done, if indeed the political will could be found and the social costs to farmers and rural towns reduced with economic incentives? The water conservancy district that manages the pumping is already taxing the water users and buying out wells close to the river, but our analysis showed that the gains for sustaining the river will be slight. Farmers could switch to crops that use less water, but corn prices are at an all-time high as I write this, owing to demand on the world market, driving even more pumping. Would you forgo profits that could pay off your farm, or help send your kids to college?

Agronomists at my university tell me that they have developed the varieties and technology needed to make similar profits with dryland corn and other crops. However, farmers are proud of the tall, green, healthy-looking corn they can grow with irrigation and are unlikely to settle for the much lower and more variable yields from dryland corn. Technology to optimize water use is just being tested, and varieties of corn that use less water are on the horizon. From our perspective as fish conservation biologists, the best option for a sustainable future for plains fishes might be to buy out wells along one representative segment of a plains river that can support these species and to simultaneously develop economically viable cropping patterns for the other basins that use less water and gradually allow some parts of the rivers to return over the next century. Otherwise, we are resigned to seeing these rivers, and the fish populations they support, extirpated during our lifetimes, and hoping that both persist somewhere else in North America. Nevertheless, those rivers and fish populations will be different, and won't reproduce what once lived in the Arikaree River and other eastern Colorado plains streams.

We eat a quick dinner and fill gas tanks in Yuma and start home toward the failing light in the west. Horizontal wedges of clouds turn shades of pink, mauve, and finally black. It will be too dark to see the Rocky Mountains when the peaks push above the horizon, though each time I see them in the light I think of the pioneers who must have rejoiced when they came in view from their covered wagons. I ponder all we learned in our research, what my students learned during this long field trip, and the interconnected fabric of humans with nature. I consider the deep interest that Shigeru Nakano and I shared about understanding these interconnections well enough to offer options for supporting both. Our headlights illuminate a swath of the shortgrass prairie that once sustained bison, passenger pigeons, Native Americans, and even plains fishes, as we hurtle into the dusk and back to my home next to the mountains. I can only hope it will continue to sustain us.

Chapter 7
Natives of the West

I guess I never really got used to working through the night. We'd drive from the last stream to the next one during bright Colorado summer afternoons, up the long dirt roads winding alongside crystal mountain streams, and past beaver meadows where moose strode and dipped in the deep pools. On the final slow miles the four-wheel-drive truck swayed and bumped along terrible roads, and we wondered whether we'd get through each rut and mud hole, or get "high-centered" on the next boulder or log. Finally, at the end, was the trailhead, where hikers and hunters left their machines and ventured into far country on foot.

Dinner was cooked and served on the tailgate in an open field or dusty parking lot, sometimes a sumptuous meal partly prepared at home, sometimes just sandwiches made on the cooler lid. Then as sunlight filtered sideways through spruce and fir crowded at the edge of the clearing and emerging insects began their evening swarms above the stream, we'd stuff tents and sleeping bags and breakfast into our backpacks, topped by warm clothes and rain gear for the cold night ahead. Boots went on sore feet and were laced up with sore hands. Hats were found, backpacks shouldered, the truck locked and the keys safely stowed, and stiff legs found the trail for the hike in.

Native cutthroat trout in the southern Rockies are difficult to get to. They tend to live in places where humans don't go much, miles beyond the stream reaches where most people choose to hike and fish. These trout subsist today primarily in small streams, too narrow and shallow to interest the average trout angler, up past waterfalls or other barriers, too brushy for good fly fishing. The trail often peters out before it reaches them, and biologists are left picking their way over downed trees along unused tracks, skirting slopes of jumbled talus, and sometimes walking through the stream itself. And often, there are not that many fish in these small streams. Indeed, in all of Colorado there are not that many streams left where the native trout, which arrived as the glaciers left, occur in abundance.

Dusk settles in the forest around us as we hike carefully along the trail next to Cabin Creek, one of our study streams, placing feet only in places where they won't slip and turn an ankle or twist a knee. At the limits of my endurance on an uphill climb, I stop to rest and breathe, bent over, hands on thighs, wiping sweat from my brow and neck, and reach for my water. My technician is younger and more fit and wonders why my breaks are so frequent. She starts ahead before I am quite recovered. Finally, we reach the section on this stream that we chose for study early in the summer and stop to make a simple camp. Kicking away sticks and cones from the few moderately flat places on the narrow bench perched between the boulders on the valley wall and the riparian willows, we pitch our one-person tents and lay out sleeping bags. We'll want to fall onto them and sleep right away when we're finished, especially if it rains as it often does. The work will be hard—kneeling, bending, crawling, slipping, and counting.

The amazing truth is that these beautiful fish, originally the most widely distributed native trout in North America, once thrived in vast abundance far downstream from here in the large rivers of the West. Like grizzly bears and elk and wolves, they ranged out into valleys and onto the plains away from the mountains, living in the wide channels that flowed clean and clear over gravel riffles and created deep pools undercut beneath huge cottonwoods in the riparian gallery forest. Their shape and hues fit them perfectly in the bright waters, concealing them from the eagles and osprey, the mink and bears that hunted them for food.

To cradle a cutthroat trout is to hold a broad iridescent ribbon of pure gold. Like most of the forms, the Colorado River cutthroat trout we find in Cabin Creek are graced on their sides with a palette grading from brassy green on top, through golden yellow along their sides, to orange or bright crimson along the belly and fins. They were known by early settlers as black-spotted trout, because their sides are sprinkled with large black spots that increase toward the tail, the largest of which are larger than the pupil of their eyes for this form. A natural icon of the West, cutthroat trout were worthy of records in the journals of early European explorers. Francisco de Coronado first encountered them in 1541 in the Pecos River south of Santa Fe, New Mexico, and Lewis and Clark found them in 1805 in the Missouri River at Great Falls, Montana. We know little about their legends among Native Americans.

As I lay down to rest for a few minutes on my sleeping bag before the long night of work ahead, I ponder, "How did we get here, with our native trout only in short fragments of headwater streams like this one?" From long study I know the answers, and that the reasons for the answers lie in myriad human aspirations and actions and consequences. At the edge of what is now Rocky Mountain National Park near Estes Park, Colorado, one market fishermen in the fall of 1875 caught more than seven hundred cutthroat trout in three days from one beaver pond and sent them by wagon to be sold in the Denver market. We proceeded from there.... Just then, my thoughts are scattered by the pit and pat of light rain on my tent fly. Twilight has faded to total darkness all around, and the rain is angled by a light wind. I sit partly sheltered at the edge of my tent, pull on my waders and lace up my wading shoes, and then roll out and spring to my feet, huddling my back to the breeze to zip my raincoat against the wet and cold. It's time to go count the spiders along Cabin Creek.

In addition to cutthroat trout, there are several other species of native trout and charr that add to the beauty and diversity of salmonids in the West. Dark green-gray bull charr are native to the great Pacific Northwest of the United States and Canada, and the Dolly Varden charr that we studied in Japan extend east across the Bering Strait and south along the coast to Puget Sound in Washington state. Various forms of rainbow trout occur all along the Pacific coast from Mexico north to Alaska, and inland to Idaho and Nevada. Gila and Apache trout are found in isolated drainages to the south and west in the mountains of Arizona and New Mexico. But cutthroat trout are by far the most widespread and occur in a great variety of forms, most of which scientists have named as different subspecies. They originally occurred all along the Pacific coast from northern California to the Kenai Peninsula in Alaska and throughout the Rocky Mountains and Great Basin of the inland West from southeastern New Mexico to southern Alberta (figure 14). They evolved in the Columbia River basin and diverged from a common ancestor with rainbow trout (the species to which they are most closely related) about 2.5 million years ago.

How did cutthroat trout become distributed throughout the West, and why are there so many different forms? This question was the life work of my late colleague Bob Behnke, a professor at Colorado State University who, over a career spanning more than fifty years, assembled and refined a systematic classification of the trout in the West. Bob examined all the various forms of native trout, including museum specimens of ones now extinct. Like a detective, he pored over data on everything from their genetics to the geological history of western North America to sleuth out how they are related and why they occur where they do. What he surmised from all this information is that, through long eons of time, the ranges of trout and salmon have ebbed and flowed in the western United States. Fossils of ancient trout now long extinct show that these fish extended down as far as southwestern Mexico during a period of cool climate 5 to 10 million years ago, and that a species of a primitive salmonid now found only in Eurasia once lived in Idaho. But the forms that gave rise to rainbow and cutthroat trout survived and continued to evolve along the Pacific coast and in the Columbia River basin of the Pacific Northwest.

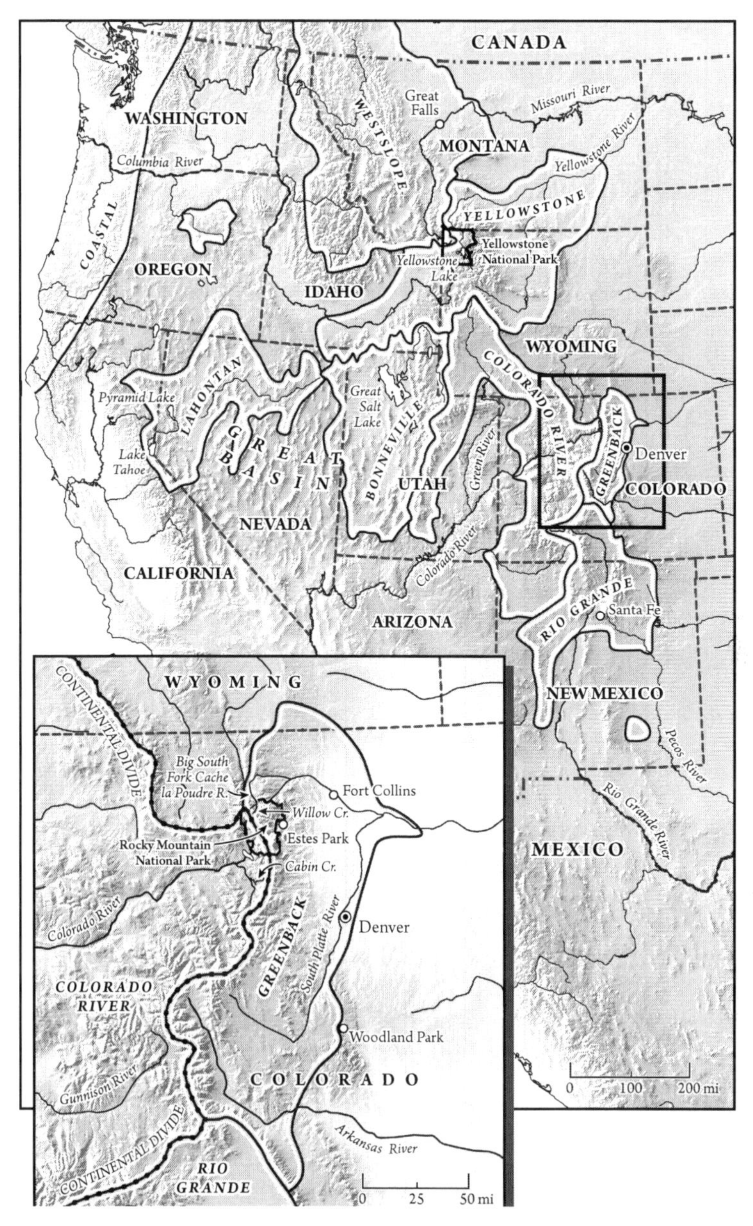

Figure 14. The original native ranges of eight main subspecies of cutthroat trout within the four main groups native to the western United States: coastal, westslope, Lahontan and four related subspecies (all enclosed within the boundary shown), and Yellowstone and six related subspecies (of these, Bonneville, Colorado River, greenback, and Rio Grande are shown). Classification and distribution are after Behnke (2002). An isolated enclave of westslope cutthroat trout occurs in eastern Oregon, and other small isolated populations in north-central Washington state and southern British Columbia are not shown. An isolated enclave of Rio Grande cutthroat trout occurs in southeastern New Mexico. The inset shows study streams and locations in the mountains of central Colorado.

About 1 million years ago, in the middle of the great series of ice ages (Pleistocene Epoch), the cutthroat trout family tree diverged into four main lines. When Euroamericans first explored the West they found these fish distributed along the Pacific coast (the coastal cutthroat trout subspecies), in the northern Rocky Mountains (westslope cutthroat trout), in the Great Basin centered on Nevada (Lahontan cutthroat trout and four related subspecies), and throughout the central and southern Rocky Mountains (Yellowstone cutthroat trout and six related subspecies). The different forms range from the large silvery coastal cutthroat trout that run from the ocean into coastal rivers of the rainforests in southeast Alaska to the small colorful Rio Grande cutthroat trout that inhabit the mountain headwaters of the Rio Grande River and its tributaries in southern Colorado and New Mexico. The largest of the fourteen different subspecies was the Lahontan cutthroat trout, once native to Pyramid Lake and Lake Tahoe, Nevada, which apparently reached sixty-two pounds. Early explorer John C. Frémont remarked that these fish were not only extraordinary in size but superior in flavor, when Native Americans brought them to his camp at Pyramid Lake in January 1844. Sadly, this unique population that grew up to 4 feet long is now gone.

Compared with the 4.5-billion-year age of the Earth, or even the last 500 million years during which fishes have evolved, most events that led to the historic distribution of the fourteen subspecies of cutthroat trout throughout the West are like a fleeting moment within a long lifetime of experience. A child laughs. An autumn leaf falls from a tree. In only the last 70,000 years or so, during and after the last glacial period within the Pleistocene Epoch, huge lakes formed during wetter periods and huge floods occurred as other lakes broke through ice dams formed by glaciers. These lakes and floods apparently allowed Lahontan and westslope cutthroat trout to move throughout large expanses of the Great Basin and eastern Oregon and Washington, only to become isolated later in the coldwater streams and rivers that remained when the climate warmed and the lakes dried. Glaciers in the Rocky Mountains created new connections across drainage divides as they alternately grew and melted, and lava flows diverted rivers into new courses, allowing the Yellowstone cutthroat trout access into the Bonneville basin in Utah and into watersheds throughout the southern Rocky Mountains.

Those transfers of the original forms of cutthroat trout into new drainage basins where they become isolated were the spark that ignited further evolution. Just like Darwin's finches, which became isolated on different islands in the Galapagos and gradually diverged into fifteen different species, the cutthroat trout of the West became isolated in many different drainage basins and were shaped by the unique climate and conditions they encountered. Those in Pyramid Lake found an abundant source of other fish for food and evolved into large predators. In contrast, the ancestral forms of Yellowstone cutthroat trout that entered the Green River at the northern edge of the Colorado River basin in Wyoming were subsequently selected for traits that increased their survival in the streams and rivers that drain the southern Rocky Mountains. These immigrants gave rise to the Colorado River, greenback, and Rio Grande cutthroat trout subspecies that inhabit the major river basins in Colorado and New Mexico.

‹ ›

It is among the great tragedies of the West that these fish are now gone from most of our rivers. Evidence that they were once great forces of nature lies in the fading ink of explorer's journals and in still-vivid memories for some of us who have lived our lives here. The 16- to 23-inch westslope cutthroat trout that Lewis and Clark's Corps of Discovery caught easily below the Great Falls of the Missouri River in northwest Montana in June 1805 are no longer there, long ago replaced by nonnative brown and rainbow trout. Likewise, the almost unfathomable spawning runs of 25,000 to 50,000 Yellowstone cutthroat trout averaging 16 inches long that entered several of the larger tributaries to Yellowstone Lake every year in the mid-1980s have been reduced to a few hundred fish or less. I remember them. In May 1983, fisheries students and I were able to reach into the edge of the Yellowstone River at LeHardy Rapids to grasp some of these beautiful spawning fish with our bare hands, during a field trip that Bob Behnke and I led to northern Rockies. These fish also attracted hundreds to thousands of bears, otters, eagles, pelicans, and other fish-eating birds and mammals. These native predators have had to go elsewhere to find food or have simply declined in numbers themselves.

Why have these ribbons of gold, these trout of vast abundance and unparalleled beauty, been lost from so many places in the West? Certainly, overfishing can reduce populations drastically, especially of cutthroat trout which are, frankly, gullible when it comes to angling. Indeed, they are the perfect quarry for kids and novice anglers who want to learn about trout fishing. However, populations reduced by angling can rebound quickly when modern fishing regulations are applied by fisheries scientists, so this alone cannot explain the abject loss of this species from so many rivers.

Unfortunately, European settlers brought much more destructive forces than angling to the West in the late 1800s. For example, in Colorado, mining operations funneled large amounts of sand and silt into streams that clogged spawning gravel, and some of this sediment carried heavy metals like cadmium and zinc that are highly toxic to fish eggs and larvae. The millions of railroad ties needed to build the transcontinental railroad across the West in the 1860s were cut from riparian forests along small streams, as were logs used for mining timbers and building towns. Many streams throughout the West were cleared of their natural logs and boulders to allow sluicing these saw logs downstream in the early summer snowmelt runoff, and this destroyed most of the refuge pools needed by trout to survive the low flows in late summer through winter (see chapter 3). Sawdust from sawmills along other streams smothered all trout eggs and insect life. Overgrazing and trampling of stream banks by free-ranging cattle also brought sediment into streams that clogged gravels and filled pools, and diversions of water for irrigation and to supply towns and cities dried entire stream sections nearly completely.

As habitat was destroyed and cutthroat trout disappeared people began to clamor for fish, and other trout were stocked to replace those that had been lost. Starting in the 1850s, fish culturists in the Midwest and eastern United States developed methods to gently squeeze eggs from female trout, fertilize them with "milt" (sperm) squeezed from males, incubate and hatch the eggs in hatcheries, and rear the fry for stocking. This led to widespread rearing and stocking of trout throughout the West, not only of cutthroat trout, but also nonnative brook trout from the East, rainbow trout from the Pacific Coast, and brown trout from Europe. For example, in Colorado, the first reports of culturing and stocking cutthroat trout and nonnative brook trout are from 1872 to 1875, and during 1885 to 1953 over 750 million nonnative brook, rainbow, and brown trout were stocked in Colorado waters. At the same time, large numbers of cutthroat trout were also being raised in other western states, especially Yellowstone cutthroat trout because of easy access to the huge runs of spawning fish from Yellowstone Lake. These fish were sent by railroad and stocked around the West, and indeed, around the world. Early fisheries scientists had no idea that the fish they were raising and stocking were a different form of the same cutthroat trout species, nor that each form was best adapted to survive in their local environment. Nor did they understand that the other nonnative trout they were stocking could affect their local native trout.

So, where do cutthroat trout in the West stand now? Two of the fourteen forms are extinct, and several others became so rare by the 1940s that they were also believed to be gone. Five rare forms are now protected by the Endangered Species Act as threatened subspecies—the greenback, Paiute, and three forms of Lahontan cutthroat trout (two are not yet formally described or named by scientists). As a result of the massive amounts of habitat destroyed, and the invasion by nonnative trout, most of the remaining subspecies of the inland West are now found as pure populations in only 15 percent or less of their historic native ranges. And, the majority of these are isolated in small headwater stream fragments above barriers that prevent invasion by nonnative trout.

Although brook trout are among the most common invaders, rainbow trout are the most dangerous, because they can hybridize with cutthroat trout and create "mongrel" trout (often called "cuttbows") that are not as well adapted to the native habitat. Likewise, the many Yellowstone cutthroat trout stocked throughout the West hybridized with local forms, so many more populations have been found to be genetically altered by these legacies of stocking. Fish conservation biologists are working diligently to conserve and restore habitat for all the subspecies, often by excluding or carefully managing grazing, logging, mining, and water diversions. Where habitat can be protected, they often attempt to reintroduce native forms to streams and lakes after removing the nonnative trout or genetically hybridized cutthroat trout. These good efforts by many dedicated fisheries biologists are keeping the remaining subspecies from sliding closer to extinction.

Like almost every case in which humans degrade nature, and every challenge in one's own life, there is rarely only one cause, or only one problem that must be fixed. So it is for the decline of the many cutthroat trout subspecies in the West. In the end, we can never know whether habitat destruction or invasions by nonnative trout were the more important factor in their loss. We do know that for most forms, what habitat destruction didn't ruin, nonnative trout helped to finish. Now, most of the streams and lakes that support native cutthroat trout have some measure of protection, either because they are located on federal lands, or through long-term easements on private lands arranged by state fisheries agencies or nonprofit organizations like The Nature Conservancy. But nonnative trout are still invading upstream in many places, and in others, people continue to introduce them illegally. For example, the incredibly bountiful spawning runs of cutthroat trout in Yellowstone Lake fell victim to voracious nonnative lake trout predators that were illegally introduced in the late 1980s and expanded their population in the 1990s. The nonnative whirling disease parasite that was introduced to the lake from an unknown source has also contributed to their decline, and may yet be the crowning blow.

∽

As we break out of the timber and seek the trail hidden in the low willows along the stream, the ache has started to leave my legs from the short but steep climb over the low divide into Willow Creek. We stop to put on rain pants to shield against the heavy dew and adjust our awkward loads of gear packed in for the day's research sampling. I shiver as the perspiration on my skin quickly grows cold in this crisp dry early-morning air above 10,000 feet, feeling more like mid-October than early September. Yet I love these mornings hiking into streams with my students to earn hard-won data that we hope will answer questions important for the conservation of these native trout. Just how is it that brook trout invade, and how do they replace cutthroat trout so quickly in these streams?

Willow Creek gathers its summer base flow from springs and spongy moss seeps that trickle from the lush subalpine meadow that coats this broad valley. The stream flows north beyond the border of Rocky Mountain National Park to join what is known as the Big South Fork of the Cache la Poudre River, near the trailhead about a two-hour drive from my home. As we hike the last two miles upstream to our study reach, the fen of low willow glistens with dew, the leaves now arrayed in hues from yellow-green to gold and burnt orange. Some flutter into the stream and swirl and eddy away downstream, the flash of their silver undersides in the deep pools looking remarkably like young trout. They will trick us into netting them later when we are electrofishing. But every so often our eyes force us to stop hiking. We are struck by the panoramic vista of the dark spruce-fir forests beyond the bright meadow, their fingers extending upward to grasp the shoulders of peaks that reach beyond timberline, each graced by slips of snow that never melted from their north faces. And from these forests beyond the meadow sound eerie flute-like tones, as though Pan haunted this far wilderness. It is fall, and the bull elk are bugling as they seek to defend their mates.

When we arrive at our field site, crews are assigned and we quickly set to the work of electrofishing. The sun rises high over the broad meadow, and the morning slips away as we focus intently on the tasks at hand. "Here's the rest of the fish from the first pass," says Doug Peterson, the PhD student leading this research project, who has returned from the electrofishing crew carrying a bucket brimming with cutthroat and brook trout for us to measure and weigh. We will also mark the cutthroat with colored dye and tags to allow identifying each individual in future years. Although we have worked hard and fast processing fish the entire morning until noon, sandwiches half eaten will have to wait because the fish won't. We must quickly finish the cutthroat trout before the water in the buckets warms and they use up all the oxygen, and place them in live baskets in the stream to recover. The brook trout are euthanized by overdosing them in anesthetic and then preserved for analysis, in the laboratory, of their age, sex, and maturity. It will take three days to complete the two electrofishing passes of the long segment of Willow Creek that Doug is studying for his four-year dissertation research project. Then we'll move on to the other three streams and do the same.

It seems odd to be killing beautiful brook trout, a fish revered by anglers 1,000 miles to the east in its own native range. But here in Colorado, the forest-green fish are invaders, first brought to Colorado and propagated soon after the railroad reached Denver in 1870. After the native cutthroat trout were decimated by early settlers, brook trout were one of the first species stocked to attempt to restore populations of trout for food and sport. They were being raised in Colorado by 1872, and 100,000 were stocked in mountain streams near Woodland Park in 1874. The native "black-spotted" trout proved difficult to propagate, whereas brook trout had been successfully reared in hatcheries in New York and Michigan, and their fertilized eggs could be easily shipped on ice by railroad from these sources. An adaptable and prolific fish, they soon were widely stocked in lakes and streams and made their own way into many of the farthest headwaters.

Evidence from the past reveals that brook trout are a potent invader once they reach streams with cutthroat trout. Bob Behnke, who in the 1960s began spurring efforts to restore greenback cutthroat trout to streams of their native range east of the Continental Divide in Colorado, reported that once brook trout gained access to streams, the cutthroat trout were virtually gone within five years. This pattern repeated itself in most streams throughout the region when brook trout invaded. To prevent the extinction of greenbacks, fisheries managers built barriers, usually small dams, near the downstream ends of headwater streams, and used one of several chemicals to remove the nonnative brook trout from upstream. But in many cases either a few fish were able to survive the chemical treatment in beaver ponds or spring seeps, or the barrier failed to keep all fish out, and the brook trout invaded again, each time taking over in about five years. How does this happen, and when during their lives are cutthroat trout most vulnerable? Doug and I wanted to find out, hoping that knowing the answers would allow more effective management to conserve native cutthroat trout in the southern Rockies.

So, as for our work in Japan, we decided to carry out a large-scale field experiment. After searching many locations, Doug found segments on four streams where cutthroat trout were temporarily holding their own and were still about equal in number to the brook trout invading upstream. In two "treatment" streams we removed brook trout for four years in a row from study segments that averaged about 1,100 yards long and predicted that survival of the cutthroat trout we marked and released would increase, at least for some age classes. In the other two "control" streams we simply measured and marked all cutthroat and brook trout in the study reaches and released them and predicted that survival of at least some age classes of cutthroat trout would be low. The streams were arranged in two treatment-control pairs, one pair at mid-elevation (about 8,500 feet) and the other pair at high elevation (10,500 feet; Willow Creek was the high-elevation treatment stream). Doug also installed fish traps each summer to find out whether invading brook trout forced cutthroat trout to leave, and to measure how many brook trout immigrated into the study reaches.

The work was grueling, no doubt about it. Wading up streams with a thirty-five-pound backpack electrofishing unit on your back, bending and probing beneath brushy stream banks to capture and net fish, stumbling and slipping along over slick cobble and boulder streambeds is real work. Add either rain or snow and it becomes something to be endured, and makes it even more important to coax the crew onward with promises of a sumptuous hot meal at the end of the day. The reward for two or three hours of carrying the backpack "shocker" is to work on the processing crew, sometimes huddling in the cold rain to measure, weigh, and tag hundreds of trout per hour, up to your elbows in cold water. In between batches of fish, it is a relief to warm up by carrying the buckets along hummocky banks to the live baskets in the stream where the tagged fish can recover. Even recording the data becomes challenging when our semi-waterproof paper eventually succumbs to the cold drizzle that seeps into everything.

Don't get me wrong. I love everything about the beautiful fish we work with and the challenge of the work itself, right down to the willow twigs and spruce needles that fall down your back when you lunge to net fish that flash deep beneath undercut stream banks. (We say that you didn't have a good day electrofishing unless your arms are scratched up and your underwear is full of these gifts from the forest.) But there are few things that have brought me to a more complete state of fatigue than a day of stream electrofishing.

After four summers of field research and four winters of detailed statistical analyses, the results revealed by Doug Peterson's work were striking. During their first year of life, cutthroat trout fry in the mid-elevation treatment stream where brook trout were removed survived more than thirteen times better than those in the paired control stream where brook trout remained, and survival of yearling cutthroat was more than twice as high. For their part, brook trout fry in the control stream survived ten times better than the cutthroat fry. More surprising, however, was that adult cutthroat trout, those two years old and older, survived just as well as the adult brook trout. The experiment proved that brook trout decimated cutthroat trout, but only during the first two years of life. And, it also gave a flash of insight into something that several of us had observed but hadn't made sense of before. In other streams with barriers to prevent brook trout invasion, we often found adult cutthroat trout that had moved downstream over the barrier were living fat and apparently happy among the many brook trout. Now we knew why.

The data from the high-elevation pair of streams added more to the story. There, the cutthroat trout were zombies, the swimming dead. The remnant cutthroat trout populations in both of these small cold headwater streams above 10,000 feet produced very few or no young in four years. Although the adult cutthroat trout in such streams may live for up to a decade, the brook trout have them backed into a corner. They rarely or never reproduce, and when they die out completely, the populations will be extirpated. Two other four-year research projects in our lab by PhD students Amy Harig and Mark Coleman revealed the reasons why cutthroat failed to reproduce in these streams. In short, if water temperatures average much below 50°F during the warmest month of the summer (July or August), when cutthroat trout fry have just emerged from the gravel and are beginning to feed, they cannot grow large enough and store enough fat to have a good chance of surviving through the winter. Water temperatures during winter are harsh, hovering near the freezing point for four to six months each year in these high-elevation streams, and when the snow melts the small fry must swim in frigid water to hold their positions against the swift flows during runoff. In contrast, the brook trout fry emerge earlier, in late spring, and have a longer growing season to feed as fry, so they are more likely to grow large enough at these low summer water temperatures to survive the winter.

To add insult to injury, Doug found that the brook trout just kept coming. The fish traps showed that about as many cutthroat trout entered the study reaches as left, regardless of whether we removed brook trout or not. However, large numbers of brook trout were swimming upstream into the reaches during all seasons of the year. Each summer we removed from the two treatment streams all the brook trout that we captured in the fish traps and by electrofishing. Unfortunately, Doug had to remove the traps each fall when the streams began to ice up, or risk losing them in the high flows the next spring when the snow melted. By the time he could install the traps again the next summer after flows had subsided, eight months later, the immigrating brook trout had replaced close to half of the original brook trout population that we had removed from the high-elevation treatment stream, and the entire population in the mid-elevation stream. In short, the brook trout didn't need to reproduce well in these high, cold headwaters because they sent waves of adult invaders from warmer reaches downstream. By the end of our four-year study, we realized that brook trout are the ideal invader in these Rocky Mountain streams.

Did these daunting results help us provide answers that aid cutthroat trout conservation? If brook trout kill most young cutthroat trout each year through competition and predation, and keep coming from downstream, then is there any hope of managing the brook trout invasion and conserving native cutthroat trout? The best option is to find long stream segments with natural barriers (waterfalls or dry reaches) and remove all brook trout from above the barrier, to create a refuge for the native trout. However, in many cases, there is no barrier and no site to install a suitable one. In others, biologists cannot remove brook trout completely because the habitat is too complex. For example, in streams with large beaver ponds, a few brook trout typically survive chemical treatments by finding spring seeps. Nevertheless, if the cutthroat trout populations in these streams are highly valued for conservation, biologists may mount annual electrofishing campaigns to remove the waves of immigrating brook trout and keep the invaders at bay.

Doug's research showed clearly that the only chink in the armor of invading brook trout is that *adult* cutthroat can survive their advance. A further detailed analysis revealed that the best strategy for managers is to remove brook trout for at least three years in a row to allow several age classes of cutthroat trout fry to survive to adulthood, followed by no more than two years without brook trout suppression. This simple rule of thumb will reduce the cost of management and yet help sustain these unique cutthroat trout populations and preserve options for the future.

The study of nonnative trout invaders was the subject of my first research in graduate school and has remained a passion for more than thirty-five years. The best of these studies revealed answers, sometimes amazing answers, and yet each answer led us to deeper questions. Scientists are never satisfied with only the first answers, nor should they be. "What are the ultimate questions?" they wonder. And, what are the answers to these deepest puzzles?

One of the deepest and most puzzling questions for the larger field that we ecologists call "invasion biology" is why nonnative species can invade in the first place. This is such a simple question, but reveals a profound paradox. After all, Charles Darwin taught us that natural selection in any environment, from the Galapagos Islands that he studied to trout streams in the Rockies, drives "survival of the fittest." So, if only the fittest survive, that means the native trout that evolved in a given watershed should be the fittest species possible, and that any introduced trout species should be inferior and unable to invade.

Well, unfortunately, the way we normally think about Darwin's theory doesn't include what happens when nonnative species are introduced. Like his contemporary Alfred Russel Wallace, who came up with nearly the identical theory at the same time, Darwin studied undisturbed tropical islands, nearly perfect natural laboratories where the products of evolution by natural selection could be studied in their untrammeled state. In this scenario, if a genetic mutation caused a small change in one member of a bird population that allowed it greater competitive ability and survival, and it passed these benefits on to its offspring, then more of these offspring would survive than those of other parents, and the inferior forms would eventually die out.

But even in Darwin's day, nonnative plants and animals had been transported throughout the more developed world by trading ships and had begun to invade. For example, common carp were brought by monks from their native Europe to England by the early 1500s and subsequently spread throughout most of the southern half of the country. After World War II, international commerce exploded, and cargo was sent by ships and planes to all the far reaches of the Earth, but the cargo often carried nonnative species in the form of seeds, roots, larvae, and adult organisms. Some introductions were intentional, carried by immigrants who wanted to bring familiar plants and animals to their new homelands. Many others were not intended, the result of hitchhiking invaders, including rats and snakes that reached tropical islands. Whether intentional or not, the introduction of nonnative species means that evolution by natural selection is no longer the only way that new organisms appear in any biological community and challenge the fittest for survival.

Nonnative trout have most often been introduced to new countries on purpose. Brown trout were sent from England (where they are native) to Tasmania and then New Zealand in the 1860s, and from Germany to North America in the early 1880s, as soon as fish culturists learned that fertilized eggs were hardy enough to allow shipping them on ice to delay their development and hatching. Likewise, during the same period, rainbow trout from California were shipped across North America and around the world, such as to Japan in 1877 and England in 1884. Brook trout soon followed, from east to west in North America, and then to about fifty other countries. Rainbow trout, the most widely introduced salmonid, have been introduced to all continents except Antarctica, and to about a hundred other countries, more than any other fish species except common carp and grass carp.

So if Darwin's theory is true, why are brook trout apparently better adapted to conditions in the Rocky Mountains of Colorado than the cutthroat trout that are native there? Why should a newcomer be able to survive better than the native trout that have evolved for the last 10,000 to 50,000 years in these streams? And this is not the only case of this paradox of trout invasions. Brook trout themselves are native to streams of the southern Appalachian Mountains in Tennessee and North Carolina, and yet rainbow trout from the Pacific Coast have invaded there, and pushed the brook trout far up into tiny steep headwater streams. I puzzled over this fundamental question for more than two decades, until the answers revealed by other research finally allowed some of the pieces to fall in place.

Our own lives seem to be shaped by happenstance. You meet someone new, or an accident happens, and life changes forever. So like this, the answer to the paradox of these trout invasions lies in quirks of fate, and legacies of evolution. The ancestors of rainbow and cutthroat trout evolved along the Pacific coast, where winter rains bring swirling floods that signal fish to migrate upstream to spawn. The female trout dig nests in gravel at the tails of pools in late winter, bury their eggs, and the eggs incubate until they hatch in early spring. By the time the fry wriggle free from the gravel in late spring, flows are subsiding and the water is warming, creating a stable and productive environment for the young fish to begin feeding and growing. It should have been no surprise, then, when we discovered that rainbow trout were most successful at invading in new regions where the natural flow regime in rivers was similar to those on our Pacific Coast, with floods from winter rains that declined to summer low flows. Streams of the southern Appalachian Mountains were one of those places.

In sharp contrast, brook trout originally evolved in the upper Midwest and Northeast of North America and spawn during fall. Stream flows are typically low and stable in this region during the winter when brook trout eggs incubate in the gravel. The fry emerge in mid- to late April after the March freshets caused by melting snow have subsided in most years. So, like rainbow and cutthroat trout in their native ranges, these brook trout fry emerge into an environment with stable flows and warming water that produce abundant tiny invertebrates for them to eat.

But the quirk of fate is that native cutthroat trout in the West, and brook trout in the East, both eventually extended their distributions southward in North America to regions where the season of flooding was markedly different than where they had originally evolved. And, by an even stranger chance, the nonnative trout we introduced to these regions were pre-adapted to survive better in these flow regimes than were the native trout. In the West, cutthroat trout moved inland and south over the last million years, eventually reaching streams in the mountains of Colorado and New Mexico. Here, melting snow produces floods of ice-cold water in early summer, just the opposite of the winter floods in the Pacific Northwest where cutthroat trout first evolved. So, inland cutthroat trout must delay their spawning until these floods begin to subside and waters warm to about 42°F, usually in mid-June. However, by the time their eggs hatch and the fry emerge in July or August in these cold mountain streams, there is little growing season left before temperatures drop to near freezing in late September. But by chance, brook trout are much better adapted to these conditions, because they spawn in the fall and their eggs incubate throughout the stable flows that last all winter. The eggs hatch and the fry emerge in spring, and by the time the snowmelt runoff comes in June at least some young brook trout are large enough to hide along the edges and avoid the brunt of the flood. Their earlier emergence gives them a much longer summer growing season, so they are much larger when they enter the harsh conditions of winter.

Almost the same quirk of fate occurred for brook trout, which moved south along the Appalachian Mountain chain over even longer eons of time than for cutthroat trout. However, streams in the mountains of Tennessee, North Carolina, and northern Georgia are subject to winter rains that produce floods, just the opposite of the relatively stable winter flow regime where brook trout evolved. In these southern Appalachian streams, eggs or newly hatched fry of native brook trout can be scoured from the gravel by winter floods and lost downstream. In contrast, introduced rainbow trout spawn in early spring, after the winter floods, and so are much better suited to this flow regime than the native brook trout.

No one ever said that evolution was perfect. We *Homo sapiens* suffer appendicitis and lower back problems because we are not perfectly evolved to fit our environment, even though we were apparently superior to *Homo neanderthalensis* at surviving and dominating on Earth. In addition, whatever unique combinations of genes allowed Neanderthals to survive in Eurasia before we modern humans arrived from Africa could not be changed quickly enough by natural selection to allow them to withstand our invasion. In a similar way, the early summer spawning of cutthroat trout is a fundamental physiological characteristic of this line of fishes, and could not be switched to fall spawning after the fish reached Colorado, or after brook trout arrived. Instead, the fall-spawning brook trout were, by chance, pre-adapted for the flow regime dominated by summer snowmelt runoff and survived better than the cutthroat trout by virtue of their evolutionary history. In addition, because brook trout fry emerge earlier in the spring, they are always larger than cutthroat trout fry, and can win in head-to-head competition for stream positions. Indeed, when cutthroat trout fry first emerge, the larger brook trout fry are already large enough to eat them.

Reaching a logical explanation for the paradox of trout invasions in North America was among the most satisfying accomplishments in my research career. In fact, when the pieces finally fell into place, and conferring with some of my closest colleagues proved that the logic was sound, I wrote the entire first draft of the manuscript in eight hours, starting at 10 p.m. and working through the summer night. It seemed like a great relief to finally record on paper the answer to this fundamental question that had been vexing me for so many years. But sometimes the most important questions in modern-day ecology are really about people. Why do humans intentionally introduce nonnative species in the first place? And why do they like them better than the native species found in that place? This is the second great paradox of invasion biology, and one that I've rarely heard ecologists discuss. It lies in the realm of sociology, or environmental psychology I guess, and we ecologists don't often cross boundaries into those fields to explore.

The first settlers in any region import nonnative species to make themselves feel at home. When I traveled to New Zealand, I found that half the plants I saw looked strange and exotic but the other half were very familiar. The English that immigrated during the mid-1800s brought dandelions and oaks, deer and rabbits, and also brown trout and Atlantic salmon from their native land. The Atlantic salmon did not survive, but the brown trout flourished and created a thriving sport fishery, all the while decimating the native trout-like galaxiid fishes of the region by their predation. Most of the native galaxiids in New Zealand now occur only in the headwaters above barriers, just like the cutthroat trout in Colorado.

A second reason that humans stocked nonnative trout was to bring something novel to a region, or replace something lost. For example, because Yellowstone National Park is on a high plateau where glaciers melted only about 6,000 years ago, fish were blocked from most of the lakes and streams by natural waterfalls. One of the early park superintendents reported that only seventeen of the 150 lakes in the park had fish. So in 1889, the U.S. Fish Commission sent 7,000 yearling brook, brown, and rainbow trout from Michigan hatcheries to stock streams in the park, and many productive and storied fisheries were created with these nonnative trout. In other parts of the West, after fish populations were decimated in many streams and lakes in the late 1800s by mining, logging, grazing, and market fishing (sometimes using dynamite), the main goal was simply to restore fish to these waters. No one knew or cared whether the nonnative fish they stocked would reproduce and spread, or affect the remaining native fish. Indeed, no one knew much about the native fish in the West, because many species and forms had not yet been described or classified by scientists. Once cultured fish from other regions were widely available and stocking became common, anglers wanted the unique fish for their region. For example, Colorado soon became famous for its rainbow trout, which later attracted dignitaries like President Dwight Eisenhower for the celebrated trout fishing in the Rocky Mountains, even though the rainbows were never a native species there.

Nowadays, many of the fish in fresh waters throughout the world are not native. The widespread transport and stocking of fish for food and angling has homogenized our fishes across the globe, just as fast-food restaurants have homogenized our foods. One can fish for rainbow trout not only on the Pacific Coast where they are native, but also in Wyoming, Virginia, New Zealand, Chile, Switzerland, and many other places where they now thrive and reproduce. In the Gunnison River of western Colorado, anglers in the early 1880s found five- to ten-pound Colorado River cutthroat trout in abundance. But when rainbow trout were brought by the railroad and stocked in 1883, the introduced trout thrived, the cutthroat trout disappeared, and by 1897 anglers were catching large rainbow trout of the same size. The introduced rainbow and brook trout have relegated cutthroat trout to the cold headwaters where they grow slowly. Few people venture that far to find them. A colleague of my father who was born in Colorado once exclaimed, "Oh, so that's why they're called 'Natives'!" when I told him that the cutthroat trout he had caught far up in the headwaters of mountain streams were the only trout species native to Colorado. It is no wonder that most people are more interested in the large nonnative brown and rainbow trout that now inhabit their local streams and rivers. For them, they are now part of their "sense of place."

So, if nonnative trout are now widespread, and most humans believe a trout is a trout, is there any reason to work hard to save the natives? After all, removing nonnative trout from streams and building barriers to keep them out is expensive. It generally runs more than six figures for the average stream, although much less than the cost of a single new highway bridge. The work is expensive because barriers that stop nonnative trout at all flows are difficult to design and construct, and streams must usually be treated several times over several years to ensure that all the nonnatives are gone. Colleagues of mine have argued in a scientific paper that restoring native cutthroat trout to much of their native range in the West is not worth the cost. They claim that the ecological, social, and economic consequences of nonnative trout replacing cutthroat trout are relatively minor and that measurable effects on ecosystem function are rare. For example, these authors argue that the ecology of brook trout is so similar to cutthroat trout that replacement of the native by the nonnative should cause little change in the flows of nutrients or energy that drive these ecosystems. Besides, most anglers can't tell one species of trout from another, and nonnative trout are widely sought after, so why attempt to keep more than a few populations of the natives around as museum pieces? Indeed, if it makes no difference to the ecosystem, can we justify the expense of doing more?

This is an important question, and one that we ecologists will continue to struggle with as many ecosystems become dominated by nonnative species. But following from Shigeru Nakano's pioneering work on linkages between streams and their riparian zones, Colden Baxter and I had a hunch that for western streams it *does* matter whether native cutthroat trout persist in these watersheds. Based on our work in Japan and our knowledge of the ecology of both cutthroat and brook trout, we suspected that the nonnative trout could have hidden effects that cross the boundary and influence riparian animals. A few papers published in the 1970s from north Idaho streams reported that brook trout ate more bottom-dwelling stream insects than native cutthroat trout. Brook trout are actually a species of charr and distantly related to the Dolly Varden that Nakano and I first studied in Japan. As such, we might expect them to feed more on insects from the streambed than cutthroat trout, which capture more insects drifting on the surface or in the water itself. Female brook trout also typically become sexually mature as yearlings, a year earlier than cutthroat trout, and so brook trout produce many more small fish with hungry mouths to feed. These juvenile fish are adept at vacuuming up bottom-dwelling insects, as well as those insects swimming toward the water surface to emerge.

This question of whether it matters that brook trout replace cutthroat trout haunted us, and we knew that attempting to study it would be a big risk. If nonnative brook trout eat more bottom-dwelling or emerging insects than the native cutthroat trout they replace, we reasoned that insect emergence should be lower from streams where brook trout had invaded than in streams where cutthroat trout persisted, and this lower insect emergence should also support fewer streamside spiders. But could we measure these differences in tiny emerging insects, and the spiders that eat them, in real streams? Our hypothesis was logical, but I think we both had our doubts.

By 2005 Colden Baxter had landed a job as a new assistant professor at Idaho State University, and we had gained more funding from the National Science Foundation for an ambitious set of field observations and another field experiment to test these ideas. So in 2006, Joe Benjamin, a PhD student from Colden's laboratory, and Colden and I and our crews went whole hog, measuring trout, emerging insects, and spiders in twenty streams. We chose ten pairs of streams spread across two large areas in Idaho and Colorado, each pair including one stream with native cutthroat trout and another similar stream where nonnative brook trout had completely replaced the cutthroat. The next year, in 2007, postdoc Fabio Lepori from Switzerland joined my laboratory, and all of us carried out an even larger field experiment than we had undertaken in Japan. We created twenty fenced sections in a southeast Idaho mountain stream in which we measured the effects of brook trout versus cutthroat trout on the entire stream food web, and on the riparian spiders fed by emerging adult aquatic insects.

( )

The light rain gathered on my raincoat and ran in tiny rivulets onto my waders and dripped into Cabin Creek below. Wading in the darkness, I probed carefully with my feet and nestled my wading shoes into the crevices between the cobbles and boulders so I didn't slip and fall. Cabin Creek was one of the streams I had chosen for our 2006 study—one where native Colorado River cutthroat trout still persisted. Bending low beneath the arching canopy of willows, I stalked the spiders with my headlamp, heron-like, thankful for the tiny droplets that beaded along each orb web and made them easy to count. Emerging insects typically penetrate only a few yards into the riparian vegetation, so most spiders attach their webs to shrubs and dead conifer branches that form the "walls" along the margins of the stream. Every tree that had fallen into the stream was also thick with spiders, attaching their webs to dead branches and stringing them horizontally over the water surface. These lucky spiders had first crack at capturing the emerging insects struggling into first flight off the shining, rippling surface. When my field technician and I finished counting them each night, the tallies in our field notebooks showed that half the ten streams we studied in Colorado had over 300 spiders in the 55-yard reaches we sampled.

Sometimes results from research are so evident that the "Aha!" moment comes early on. You can see clearly what the results show, right away. For example, Nakano and I could easily see that the Dolly Varden we studied in Poroshiri Stream shifted their feeding behavior within twenty minutes after we filtered out the drifting insects. But more often, epiphanies from ecological research reveal themselves slowly, after long sorting of samples and painstaking analysis, especially in this kind of research on stream food webs. For each of these two studies, three years of sorting samples by technicians in our two laboratories and analysis by Joe and Fabio were required before each manuscript could be written. Several more years filled with reviews and revisions by all of us passed before the papers appeared in journals for others to read.

There were times when it seemed like all our work would come to nothing. After two long summers of trudging trails, building fences, collecting samples, and electrofishing streams, it seemed like our results were weak, and the risk we had taken had not panned out. True, fewer insects had emerged through each square foot of the surfaces of streams invaded by brook trout than those with cutthroat trout, but at first we saw no difference in spiders between the streams. Then suddenly, I realized that we were analyzing the data incorrectly. Spiders don't care how much insect biomass is emerging through a square foot of stream surface. Because they live on the "walls" of the streams, they care about the *total* weight of insects emerging from the entire stream surface and flying up to get caught in their webs. Once we calculated this total for each study reach and used it in our analysis, the results became clear. More invertebrates, in total, meant more spiders. Although this seems like such a simple truth, I still remember the place and the moment when I exclaimed, "Aha! That's it!"

So, does it matter to the ecosystem whether or not brook trout replace native cutthroat trout in small mountain streams of the West? Both of our studies showed that it does. When we compared the ten pairs of streams in Colorado and Idaho, we found that the density of insects emerging from brook trout streams was only about two-thirds the amount from the paired streams with native cutthroat trout. Likewise, our field experiment the next year showed that brook trout in the fenced study reaches could reduce the density of emerging insects by more than half compared with native cutthroat trout, when the two species were at the same density and other factors were controlled. And when we analyzed the spider numbers, our predictions from both studies showed exactly the same result. The reduced total emergence from reaches with brook trout would cause one in five spiders to disappear.

Could the replacement of cutthroat trout by other nonnative trout in streams, rivers, and lakes in the West have any broader consequences? When brook trout invade and reduce the seemingly unimportant and largely hidden linkage of insect emergence to riparian zones by a third or half, will this really matter to anything other than spiders? Based on our experiment, brook trout can reduce the weight of adult aquatic insects entering the riparian zone from a typical 10-foot-wide stream by nearly thirteen pounds per mile in a hundred-day summer season. Birds like warblers and flycatchers that migrate to these forests for their four-month summer breeding season, after wintering in Central America, rely on emerging insects for nearly a third of their energy needs. These insects are critical to their survival after the long migration and supply the energy needed for males to defend territories and females to lay eggs. Loss of this much emergence would eliminate the food required by seventy-one birds per mile, which a study by Nakano and Murakami along Horonai Stream in Japan suggested would be two-thirds of the birds breeding there.

Losses to other nonnative trout are similarly daunting. When lake trout invaded Yellowstone Lake, cutthroat trout declined drastically and now few migrate into tributaries to spawn. A total of forty-two species of birds and mammals depended on these cutthroat trout for food, including grizzly and black bears, river otters, mink, bald eagles, ospreys, and American white pelicans. During the breeding season, cutthroat trout make up a quarter of the food of bald eagles, and these birds have disappeared from other waters when migrating salmonids declined. Mammals like mink, otters, and bears also transfer the nutrients from trout into the riparian zone when they deposit their waste products after meals of fish. This activity brings critical nutrients like nitrogen that are in short supply from the streams to the terrestrial environment. Research in Yellowstone Park, as well as near Alaskan streams where bears fish for salmon, has shown that this nitrogen is rapidly taken up by shrubs and trees and increases plant growth. These connections of everything to everything else are what weave the fabric of ecosystems. So, when the runs of tens of thousands of migrating cutthroat trout from Yellowstone Lake were lost to lake trout predation, a key process that drives the Yellowstone ecosystem was also lost. In contrast, lake trout spawn in deep water, do not migrate into tributaries, and are not available to fish-eating birds and mammals, so none of their nutrients can enter into the riparian ecosystem in this way.

And what of the economic and social values lost? Are people of the West any poorer because streams and lakes support nonnatives instead of native cutthroat trout? Around 1990, before lake trout invaded Yellowstone Lake, about 51,000 anglers fished in the section of the Yellowstone River downstream from the lake to catch and release the large Yellowstone cutthroat trout, a world-class fishing opportunity. Each angler spent just over $100 per day for this experience, which generated about $5.2 million dollars annually to the local economy. In addition, a third of a million non-angling visitors to Yellowstone National Park (of the 3 million visitors annually) visited two locations along the river just to view the beautiful large spawning fish. Who knows how many others stood in wonder and awe of the bears and eagles that gathered to catch fish on other tributaries? Who knows how many more would value these trout and bears and eagles, that they will now never see, just because they are a symbol of a way of life in western North America?

Where do we go from here? Must we everywhere lock native cutthroat trout away in tiny headwater streams, far from the advances of nonnative brook, brown, rainbow, and lake trout? In many places, this is the only current and realistic option. Managers are working diligently to expand these populations throughout the Rocky Mountains wherever they can. We and other researchers and managers are also analyzing which of these populations will be the most resilient to future pressures of a changing climate, and the warming, drying, and increases in diseases and other invasions that this will bring.

But in other places there are opportunities to offer these beautiful fish, these legacies of the West as it was when Lewis and Clark arrived, to the public for experiencing, viewing, and catching and holding in hand. For example, brook trout apparently don't invade well in the northern Rockies of Idaho and Montana, probably in part because periodic rain-on-snow events create massive winter floods that scour their eggs and fry from the gravel nests. The westslope cutthroat trout that are native there may also be better able to compete with brook trout by virtue of the unique ecology and behavior of this form. Some of these rivers haven't been reached by invading nonnative rainbow trout, and in these places modern angling regulations have allowed the cutthroat trout to thrive and grow large. For example, in several major rivers in northern Idaho, westslope cutthroat trout were reduced to only small populations of small fish by 1970 because anglers were allowed to keep up to fifteen trout of any size. Cutthroat trout are quite easy to catch, so even a modest amount of fishing removed most of the large trout every year. But after regulations were changed so that all trout were released and only artificial lures could be used (which prevents most fish from swallowing hooks deeply), the number of trout increased by more than ten times in only five years. I can attest that it is now not uncommon to catch 15-inch cutthroat trout in these streams.

Western native trout, and the streams that support them, are gifts that hold for us a sense of place, and a glimpse across the long expanse of time over which they changed and evolved. But native trout can return to streams only when people view their existence as essential to their lives.

And this can only happen if we can see and touch these amazing creatures for ourselves. My colleague Bob Behnke, whose study and writing over a fifty-year career did more than any other biologist to champion the conservation of cutthroat trout in the West, pointed out that people cannot relate to endangered species that are locked away so they cannot experience them firsthand. When anyone who wants has the chance to catch and hold in wonder these ribbons of gold, these products of a million years of evolution, in the glistening shallows at the edge of a trout stream in the West, then we have a chance to gain their support for conserving the remaining forms of cutthroat trout that we have been blessed to keep.

Chapter 8
For the Love of Rivers

That the good life on any river may likewise depend on the perception of its music, and the preservation of some music to perceive, is a form of doubt not yet entertained by science.

—Aldo Leopold, Song of the Gavilan in *A Sand County Almanac*

Late summer in northwest Montana brings hot dry weather, when farmers cut their last crops of hay off the expanses of irrigated meadows, nursed by the streams and rivers after they wander away from the narrow mountain valleys. The long dirt road leads just into the canyon, to the trailhead where backpackers set off into the Bob Marshall Wilderness Complex, one of the largest in the lower forty-eight. Like Hokkaido, this is a land that still supports grizzly bears, and my colleague Bruce Rieman and I had seen their footprints and called their name when we hiked some miles into the wilderness to fish the year before. But today, I want just to see the river and to hold in my hands the native westslope cutthroat trout that still range throughout these watersheds. These fish find this wilderness river the same as when William Clark and Meriwether Lewis first reached the region more than two hundred years ago now, and it is these explorers after whom the fish was given its formal scientific name, *Oncorhynchus clarkii lewisi*.

I am awed by history as I enter this place, this canyon of the North Fork Big Blackfoot River. This is the headwaters of the river where Norman Maclean (*A River Runs Through It*) and his brother Paul came to fish for these same trout in the teens of the last century. And so it is part of the wellspring of images and experiences that inspired perhaps the most poetic work of narrative "fiction" about rivers in western North America. As Wallace Stegner wrote in a tribute, the story that Maclean wove around the river was so close to his own experience that one suspects he didn't even bother to change the names. Like most wilderness areas, this watershed was also the scene of fierce personal struggles by a few committed people over more than a decade in the 1960s to conserve it, when other interests sought more roads and trails and logging. It seems that no wilderness area is saved without great personal sacrifice by a few key people.

The trail that leads into the expanse of high-country wilderness above hugs the sparsely forested valley wall, and today the hot dry mountain winds make the trees moan as though human. But I head the opposite direction, clambering down into the riparian zone to seek a view of the river that most others simply pass by as they aim for more remote areas upstream. The descent is steep and treacherous, as I pick a path among the sharp-edged boulders of very old metamorphic rock, twisted and folded by forces of great heat and pressure in the Earth's mantle before being uplifted into these mountains. And yet the scene that greets me is worth the effort. It gives testament to why Norman Maclean was haunted by these waters.

The river is without any tinge of color, crystal clear, in its shallows, and more beautiful than any coral reef in its depths, where the water grades from pure aquamarine over boulder-strewn runs to emerald in the deep pools. The clear blue-green waters center the eye on a valley lined by scattered cottonwoods and bordered beyond by lodgepole pine and Douglas-fir climbing the slopes of the majestic mountains. Short turbulent riffles release their curling eddies onto the surfaces of long deep pools, here and there luminescent, the shimmering water obscuring the native trout from view. The cool breeze in the canyon, and the voice of the river, give a gentle reminder that places like this can provide refuge from the world beyond. And yes, at intervals a tiny hook wound with a bit of hair and feathers and drifted across the surface of a run or pool may tempt a trout to rise. And yes, even an angler as unskilled as I can be graced by the chance to hold for a brief moment a thing as exquisitely adapted for this river as a westslope cutthroat trout, large and old enough to have reproduced several times and requiring two hands to cradle gently in the water.

And climbing out of the valley at the end of my day, through a thick stand of young lodgepole pine grown back after a fire, I stop to ponder an essential question. What is it that draws me near to rivers like this, and to this river? Why do I value them? Indeed, why do I need them? Surely, it is more than the fish or the fishing, since I could not claim to be either avid or skilled at angling. And if I recognize that my need is a deeper thing than just the fish, or only the fishing, then what is the source of that need? And if that need is real, and places like this offer something that no others can fulfill, then what is it about these places that would lead me to want to conserve them?

Although I have often sought and worked in relatively undisturbed places and on relatively pristine streams and rivers, most flowing waters are in trouble worldwide. More than 40 percent of the 3.5 million miles of rivers and streams in the United States are in poor condition, primarily from dams, diversions of water, channelization, releases of treated sewage, and from logging, grazing, draining, and plowing of riparian forests, grasslands, and wetlands. Conditions are similar throughout much of the world. For example, there are more than 77,000 dams at least 6 feet high in the conterminous United States, and even the relatively small island of Hokkaido has more than 1,200 of this size, each sufficient to prevent fish movement and change the natural patterns of stream flow and temperature on which fish and the rest of the community of aquatic life depend. For aquatic ecologists like me, the penalty of our ecological education is to live alone in this world of wounds.

And what harvest have we reaped from these uses of our streams and rivers? The degradation and loss of habitat has caused the extinction and near-extinction of organisms from one of the greatest fountainheads of diverse species on the face of the Earth. Even though fresh waters make up less than 1 percent of the surface area of the planet, and make up only a hundredth of 1 percent of the entire volume of water, they support one of every seventeen species so far described by scientists. This is more than seven times the number of species we would expect based on the small fraction of area that freshwaters cover.

Freshwaters, and especially tropical rivers and lakes, are especially rich in fishes. Of the approximately 100,000 species of animals (including invertebrates) so far described in freshwater, about 13,000 are fishes, or one of every eight species. These freshwater fishes make up nearly half of all fish species (all the others are marine or estuarine), and about a quarter of all vertebrate species (including amphibians, reptiles, birds, and mammals). About three-quarters of freshwater fishes live in the tropics, especially in large rivers like the Amazon, Congo, and Mekong, and the lakes in their floodplains. Freshwater fishes are clearly important among the vertebrates, and rivers and streams are important habitats that support these fish.

But the majority of all freshwater animals are in serious trouble. In the United States alone we have among the strongest laws to prevent extinction of species compared with nearly all countries in the world. However, despite this protection, more than a third of all fish species, half of all crayfishes, and nearly three-quarters of all freshwater mussels are included on lists of imperiled species, in categories ranging from vulnerable to already extinct. This is far greater than for reptiles, birds, or mammals, for which only about a fifth are imperiled at most. Although extinctions have occurred throughout the history of life, the rate of extinctions during the last few hundred years is at least several hundred times faster than the natural rate. Paleontologists can estimate this natural rate of extinction by using fossils to determine how fast species were lost before humans dominated the Earth. These numbers suggest that we are near a tipping point. Renowned evolutionary biologist E. O. Wilson noted that the first 90 percent of habitat lost causes half the species to disappear, whereas the final 10 percent causes extinction of the other half. We are likely at this point, where losing more habitat will cause fish and other species to go extinct much more rapidly.

In the end, the numbers describing the loss of habitat provide the reasons we have lost so many species. In turn, the loss of native freshwater fishes, crayfishes, mussels, and even lowly native snails is a sensitive indicator of the degradation and loss of clean, free-flowing streams and rivers themselves. We have been unable to stem the loss of these habitats and the species they support, and this decline continues apace. And, as Aldo Leopold observed, most of the damage inflicted on these ecosystems "is quite invisible to the layman." For example, among the lay persons I know, none implore their elected officials because 40 percent of their rivers are in poor condition, or a third of the fish species are extinct or imperiled. But should we care, as aquatic ecologists who know better, and try to make the losses visible to others? Or should we pursue the only other option that Leopold offered? Should we "harden our shells," pretend that the losses our science has revealed are none of anyone's business, and keep our ecological educations to ourselves?

Like rivulets of water coalescing into small creeks, then rivers, and eventually reaching the sea, my pondering arrives at a fundamental question—Do humans really *need* rivers? When we face facts, we have not been very successful at bringing many intact rivers along with us to the present day, so is it possible that we can just get along without them? Or are they, at most, an amenity, an option we can take or leave as we like? Or, as a third option, does what we know about humans and rivers lead us to the opposite conclusion, that healthy flowing waters are essential to our survival and well-being as humans? And if they are essential, where does this need come from, and why must we have them? Is there more here than just the value of the water, monetized and amortized into the future?

All that I have studied and understood about rivers and streams, from our work in Japan and throughout the West, and all that I have thought and felt and attempted to place in some order in this book leads me to this opposite conclusion. My argument here is that, indeed, humans really *do* need intact, healthy rivers, not only for water to drink and to grow food, and fish to catch and eat (or simply cradle and admire), but also for innate reasons, some of which we do not yet fully understand. All that I have learned and experienced, and gained through intuition, leads me to argue that we are drawn to these places for deeper reasons, many beyond the reach of language. It is for these reasons that rivers can help us to be whole and healthy people. And it is in these answers that we may find reasons to conserve rivers.

At the core of our need for rivers is that we cannot survive without clean water. Almost eight of every ten gallons of freshwater we use for all our needs, from drinking to agriculture to industrial processes, comes from streams, rivers, and in some cases lakes that are fed by rivers and streams. But the rest comes from groundwater that feeds these same sources, so in essence all our water comes from streams and rivers or their water sources. Only less than half of 1 percent comes from desalinization of saltwater, mainly because it's too expensive, uses too much energy, and has other bad effects on the environment. We do use some saltwater in the process of generating electricity, but this cannot replace the freshwater we must have to stay alive and grow food.

For those of us who live in developed countries, it may be a surprise to learn that we humans are also heavily dependent on freshwater ecosystems, especially rivers, for fish and other freshwater animals as food. In fact, about one in every six people of the 7 billion of us who inhabit the Earth relies on fish as their primary source of protein. In the developing countries of Africa, 30 percent of animal protein comes from fish, and in many inland regions of Asia vast numbers of people depend on rivers for protein nourishment. Most developed countries have already lost this huge source of protein, having traded the water away for agriculture (to create protein less efficiently) and electricity and other uses. Many developing countries are racing to do the same. For example, the mighty Columbia River in the Pacific Northwest, the second largest river within the United States in amount of flow it carries, once had runs of 10 to 16 million salmon and trout returning from the ocean each year. Now only about 1 million return, and many of these were released from hatcheries. And in 1908 the Illinois River, a short 275-mile-long river that drains diagonally across the state of Illinois, yielded 10 percent of all freshwater fish caught and sold in the United States, more than any other single river except the Columbia. But by the 1980s the catch had dwindled to virtually nothing, sacrificed to pollution and habitat loss. Imagine having this original harvest of animal protein just for the effort of fishing, without plowing fields or growing crops or feeding sheep, cattle, or pigs. Imagine the plight of millions in developing countries when this food supply is lost in the rush to build dams and divert water.

But there is much more to the value of rivers. Beyond these strictly utilitarian uses of rivers and streams are values and desires that grade from the aesthetic to the restoration of our bodies and minds. Many of these needs and yearnings are based in the evolution of our brains throughout millennia in the original landscapes where we became human. Indeed, we are apparently genetically programmed to learn to seek these places to increase our own well-being, and through this our own survival. Rivers influence us psychologically and physiologically, for reasons that lie deep in our evolutionary past. This is a fascinating hypothesis, which bears further exploration.

I stand at the trailhead atop the high ridge of Cascade Head, a wild headland on the Oregon coast protected within several forest preserves. My sabbatical leave brought me to the Sitka Center for Art and Ecology nearby to work on this book, and on this uncharacteristic partly sunny day in late October I seek inspiration from these majestic surroundings. The trail descends in switchbacks that lead me ever deeper down through the old forest, toward Hart's Cove carved into the cliffs along the ocean. I pass huge columns of Sitka spruces, their lower branches wearing chartreuse mantles of mosses, each tree so large at the base and so tall that I struggle to take it all in. I am a tiny creature among these forest druids, some now fallen and toppled into the thick duff by windstorms, others now nurse logs for small hemlocks that sprout up from their decaying ribs, reaching for the new sunlit gap aloft. The forest drips as drizzle and coastal fog condense on leaves and needles, and by late afternoon when I return dusk is already marching solemnly across the hummocks of the forest floor.

As I make the long hike back uphill from watching the surf play tag with the cliffs in Hart's Cove, I stop to rest, and realize that I can hear a small stream in the distance. The trail crosses two streams that descend westward to the cliff and then freefall a hundred feet into the sea. The broad range of frequencies I hear marks the sound of water cantering over rocky riffles and cascading into small plunge pools. The lilting and laughing tones are reminiscent of family conversation not quite heard as one awakens slowly from a deep sleep. I am drawn into this conversation with streams. How can one *not* want to move closer, to see the stream, after hearing these sounds? The scientist in me wants to conduct an experiment to test how many people would choose to walk in which direction when given the choice.

Is this alluring sound part of what draws us in and connects us to rivers? Many writings, including those of the conservationist Aldo Leopold, refer to our innate preference for the music of running water. And recently, researchers in Spain, England, and Korea actually have used controlled experiments to explore this wisdom. All of these studies confirm that people preferred the sound of streams over all other natural and human-made sounds tested. And these preferences yield benefits for us as humans. The sound of running water can induce relaxation more than any human-made sounds, based on physiological measures like heart rate and muscle tension. Perhaps part of the reason is that sound connects us more intimately to water than sight, because we must take the time to immerse ourselves in it, and hear it out.

I start moving very slowly toward the sound, straining a bit in the dim light to see the rushing water. But even before glimpses come into view through the salmonberry shrubs along its banks, I can *sense* the presence of the stream and its riparian zone. The air suddenly feels cooler and more humid, even though this old-growth forest is everywhere cool and humid. This moderate microclimate is essential for many organisms that survive only here, including amphibious salamanders that live along streams and the diaphanous and ephemeral adults of aquatic insects like mayflies that emerge and then live only a few days before mating in swarms above the water. But isn't this riparian shade and cool microclimate as essential for us as humans, especially in hot, dry climates? Environmental psychologists and human ecologists propose that humans gravitated toward riparian zones during Paleolithic times for shelter from sun and wind, taking cues from the verdant vegetation and colorful flowers that the environment was fertile and would increase their survival. This innate preference for taking refuge in the cool and moist riparian continues even now when we lay out paths for walking and biking and choose vantages for Sunday picnics.

As I move closer and the stream comes into full view, I recall the powerful allure of ever-changing vistas and glimpses of streams and rivers as one narrows the distance between. In his oft-quoted work *Biophilia*, Harvard evolutionary biologist and writer E. O. Wilson first summarized the evidence from anthropologists and environmental psychologists that our attraction to natural landscapes with water features like streams is genetically programmed, a result of the habitats in which our brains and bodies evolved. In short, the argument goes that natural landscapes with three main features—open scattered trees, water courses, and promontories offering lookouts and shelter caves—provided resources critical to the survival of early humans. Humans attracted to these places, and capable of learning and remembering their locations, survived better than those that didn't, resulting in natural selection for genes that influence these behaviors.

Subsequent studies have found that these responses are indeed innate. They are assumed to be genetically based rather than learned because they are present in small children and across cultures, and occur rapidly without thinking. Evolution of these human behaviors proceeded for more than 125,000 years in the savannas of Africa before we *Homo sapiens* spread to other continents. In contrast, only in the last few hundred years have many humans occupied urban areas where water courses are degraded or lost altogether. Therefore, it stands to reason that we modern humans also would be genetically "hardwired" in certain ways to innately prefer natural landscapes that are similar to those along streams, rivers, and lakes in the savannas in which our brains evolved.

One of the most intriguing findings, by environmental psychologist Roger Ulrich, is that simply viewing natural landscapes that include trees and water can relieve stress and thereby restore health and enhance creativity. From the perspective of evolution, early humans would have increased their chances of survival by finding places to rapidly recover from the stresses they faced, such as encounters with dangerous animals or clashes with other bands of humans. These refuges would have allowed them to recharge their physical and mental capacity so they could withstand new threats and survive. Repeated experiments revealed that after people viewed scenes with scattered trees and water courses, they reported feeling more tranquil, had lower muscle tension, lower blood pressure, and felt less anger and aggression than control groups that did not view the scenes. In addition, patients recovering from medical operations ranging from dental procedures to open-heart surgery had much less anxiety and required less medication when they could view natural scenes through windows or in pictures, especially those that included natural water courses.

Ulrich goes further to suggest that natural landscapes, by reducing stress and increasing happiness (positive emotional states, in the language of psychologists), also can increase creativity and problem-solving ability. Would more access to natural landscapes, and more of them that included intact streams, increase creativity and improve decisions in business, government, and education? These studies suggest that healthy rivers in natural landscapes could be at least as important as, for example, better computers in helping promote the creativity on which our economy and future well-being depend.

But why rivers? Lakes are also important natural water features, and in some regions are very common. In fact, my first experiences with water and fish and other aquatic animals are from natural lakes in Minnesota, and they will always hold a special appeal for me. But as Oregon writer John Daniel noted, "We don't tend to ask where a lake comes from. It lies before us, contained and complete, tantalizing in its depth but not its origin. A river is a different kind of mystery, a mystery of distance and becoming, a mystery of source." Environmental psychologist Stephen Kaplan writes that this mystery, the chance that something more could be learned about a landscape if one traveled just over the hill or around the bend, is one of the most important features that humans prefer about landscapes. Rivers and streams embody this essential quality of mystery more completely than nearly any other natural environment.

When we do the math, it turns out that streams and rivers are also much easier for most of us to reach than lakes. Our colleagues, the geologists who study how river channel networks form, have calculated that it takes roughly 100,000 of the smallest streams to produce one Mississippi River. If we spread these 100,000 small streams, and the rivers they form, evenly across the entire Mississippi River basin, this means that a stream or river is within a few miles of every child and adult in the basin, and indeed in most regions of the planet. This is closer than any lake for most people.

These neighborhood streams can also contribute in essential ways to normal healthy development of our children. Who has not seen kids make small dams of rocks in their local creek, just to play and experience the water rushing over them? Richard Louv, in his widely cited book, *Last Child in the Woods*, coined the term "Nature Deficit Disorder" to describe the malaise that our cloistered children face because we prevent them from playing outside. In contrast, he reports that contact with nature reduces ADHD in children and improves the self-image and motivation of children with disabilities. Small streams are the closest aquatic environment where children can experience these life-enhancing benefits, be drawn to their sights and sounds, and wonder at the rich ecological connections and complex biodiversity of these accessible ecosystems. Most of us are not privileged to live on the shores of an ocean or a lake. But small streams or rivers are within walking distance of nearly every child, and if restored, their beauty and complexity can rival that of any coral reef or tropical rainforest, right in our own backyards.

While working for a few weeks with a colleague in the city of Girona, near the Mediterranean coast in northeast Spain, I came upon a beautiful five-tiered fountain in a park called the Jardins de la Devesa (Gardens of the Riparian Forest) on a hot summer day. Each tier was festooned with a thick ring of mosses, ferns, and other green plants, through which tiny rivulets of water cascaded down to the next level. It reminded me of a cool and inviting waterfall we had seen a few days earlier, on a small stream draining the foothills of the Pyrenees Mountains to the north near the border with France. I suddenly realized that a fountain is our human creation to mimic a waterfall and create the alluring sights and sounds of running water.

The largest artificial waterfall ever created was designed by architects Michael Arad and Peter Walker for the memorial to the victims of the

9/11 disaster, in the footprints of the two World Trade Center buildings in New York City. Perhaps innately, the architects of this beautiful memorial incorporated two main features that the evolutionary biologists and environmental psychologists found most attractive to humans about natural landscapes—a grove of scattered trees and the sounds of running water. In the falling water they found elements that reflect the endless flow of time, and in the trees whose leaves grow and die with the seasons they sought the timelessness of loss and rebirth. But I believe we also seek these sights and sounds at times when we want and need to restore our souls, to help move us from heartbreak to something bearable. And in the end, even the scientists propose that natural landscapes with rivers, and the seminatural ones we create, offer us the chance to find deeper meaning and more fulfillment in life, and perhaps can even mean the difference between sanity and madness.

But the same question keeps coming back to me, and I have struggled with it for most of three years now. Beyond the science of it, beyond even the many benefits based on our own self-interest, what is essential about rivers for us as humans? Why do we need them?

If rivers are a commodity that we conserve because they provide us benefits, even those as intangible as psychological benefits that we seek innately, then this commodity can be sold off to the highest bidder. If we can use our science to tease apart the features of sight and sound that comfort and heal us, then we probably can convince ourselves that we can artificially re-create these "ecosystem services" somewhere else. Then, who needs real rivers?

But is there not something deeper, way beyond the science of it? Isn't there something beyond even our own human aspirations, perhaps beyond the reach of language? In short, what would cause us to want to conserve rivers, even if we ourselves received no personal gain?

Leopold seems to point us to a hard road, when it comes to conservation. He writes, "We shall never achieve harmony with land, any more than we shall achieve justice or liberty for people. In these higher aspirations, the important thing is not to achieve, but to strive." This admonishment at first seems like a paradox, but continues to haunt me, and also inspire me. Why did Leopold write *never*, in the first premise? In the end, I think he realized that we humans will always treat land and its waters poorly, as long as we believe we own them and therefore have the right. Likewise, we will always treat poorly those other humans whom we consider inferior, by reason of race, background, or circumstance. In both cases we act this way simply because we have the power to do so. We fail to understand the true consequences of our actions for the future of the land, or for our human societies, and for our offspring who must live with both.

And as to Leopold's second premise, he seems to point to the importance of continuing against the odds, sometimes against rational hope of making much progress. Perhaps this is a matter of grace, or style, as philosopher Gary Snyder has remarked. My family knows this striving well, as ones who have nurtured and supported a daughter and sister with multiple disabilities. Despite climbing many mountains to learn to walk and to talk, graduate from high school, and earn a Master of Arts in Special Education, Emily was denied the future that she had prepared for. School principals would not hire her, despite her strong qualifications and value as a role model, apparently because of a mild speech impediment. Coworkers and supervisors in a nonprofit organization for adults with disabilities didn't like the way she talked and asked that she resign. The Americans with Disabilities Act offered little help, so after we all struggled in vain to seek fairness and an equal chance, she resigned in disgust. Through her own efforts, she earned a job managing an office for the U.S. Fish and Wildlife Service, and she now works to help conserve fish. Through these experiences, we all learned the true wisdom of striving for these "higher aspirations" and never giving up hope. Whether striving for harmony with land and waters or justice for people, for those who know something deeply, to do otherwise would be to give up one's soul.

And, it is with the importance of this striving that Leopold continues his essay, titled "Conservation." It must grow from within, he argues, as "an effort of the mind as well as a disturbance of the emotions." It must come from something deeply held, beyond a sole focus on cerebral arguments for utilitarian values of water or food, or even the less tangible benefits for health and creativity. It must tap something deeper even than our narrow self-interest, including that intangible legacy that E. O. Wilson and others argue we retain because our brains seek to fulfill the needs of our evolutionary past.

We know so little about this, and yet it is so critical. What *are* these emotions that cause us to want to conserve the places that sustain us as humans, those places that we care so about? To state it plainly, the greatest of these feelings is love. Aldo Leopold spent a lifetime becoming a scientist and a professor, indeed the first professor in the field of wildlife management. However, at the end of his career he realized clearly that, beyond the science that we must have to do the job, the challenge in conservation rests with the imagining and feeling part of the mind. In short, it is what most of us would call a matter of the heart. And this realization emerges in his writing, as he moves from describing the need for "a disturbance of the emotions" in his essay "Conservation," to using the word "love" outright in other essays, such as "The Land Ethic." "It is inconceivable to me that an ethical relation to land can exist without love, respect, and admiration for land, and a high regard for its value." What risks did Leopold ponder, as a professor, in using the word "love" to describe his land ethic? Would his stature as a wildlife researcher be called into question? If so, he apparently decided that the chance to consider values not yet entertained by science was worth the risk. Ultimately, he concluded that the evolution of this ethic is not only an intellectual process, but also an emotional one.

And what ethical and moral basis for conservation do these emotions lead us to? What is it about our love for these places that would lead us to treat the land and its rivers and streams with respect, and extend to them the care we now reserve for other humans and our communities? In one of her beautiful essays, environmental philosopher Kathleen Dean Moore writes about her discovery that our love for places has all the same elements as our love for other people. We want to be near them, and to know everything about them. We care about their welfare, fear their loss, and are lifted and transformed by their presence. We want to protect them, fiercely, and are helpless to do otherwise. We arrive at the form of love that in Greek is called *agape*, when we desire good things for the beloved, rather than for ourselves.

And from this love, Moore writes, a deep moral responsibility grows, an "ecological ethic of care." In this world we are embedded in relationships not only with people who care for us and about whom we care, but also with the landscapes and riverscapes that sustain and nurture us, give us water, food, shelter from the sun, and the music to comfort our loss and lift the human spirit. If we care about these relationships, then this tells us that we ought to act in ways that conserve and sustain these things that offer us a life worth living. We reach "moral ground" for conserving these places we love when we find ways to transcend the purely scientific issues. Evolutionary biologist Stephen Jay Gould had it right. "For we will not fight to save what we do not love," and we will not love those things for which we do not develop a deep, abiding, emotional bond.

But I realize that this argument must be extended even further. I argue that we are still at the point that we do not fully understand *why* we love rivers and streams, and what we would miss if they were gone. Indeed, I argue that we will not work to save what we do not yet even understand that we love. And so, what if I ponder what I love about rivers, way beyond the science of it, and hope in that love to find the reasons to conserve them? What if I take the risk to articulate what *I* love about the music of these waters, even as a scientist and a professor who has spent more than three decades taking apart the instruments of the great orchestra that plays this song of songs, to see how they are made? What about when I am about to leave this good Earth, and at the last consider what I love and hope to leave my children, and theirs? What can I offer, as reasons to love rivers?

Most of all, I say, love them for their shimmering runs and curling eddies, their riffles that speak a language that we once knew to the pools where the fish rise in search of their drifting meals. Love them for the cool heavy air that settles in the riparian at eventide after hot summer days, when one is camped far from home. Find wonder in the mists steaming from the bends, turned bright by the rising sun on crisp fall mornings, and the jeweled dew falling cold on your neck as you thread through the willows. Ponder in awe the vistas of rivers and their valleys as you encounter them anywhere in the world, and the ever-changing glimpses of streams as you hike down to greet them.

But even more, hold in reverence the animals and plants that you find there, so surprisingly colorful and intricate and complex, and so well adapted through millennia to the relentless pull of the flow, ever seeking the oceans. Find yourself drawn to the movement of the stream itself, and when you wade in, feel the raw power against your shins and thighs. But deeper even than these emotions, seek the murmuring, lilting half-voices that can calm for you the inner half-voices of doubt and fear that haunt your soul. Love the mystery only partly grasped, that the river flows here but at the same time everywhere, a gathering of waters from places beyond and unseen. And in this mystery, realize that the river is not a thing, not an object or even a place, but a journey, so like life itself.

And in this love, return to the river that connects you to itself, and to all life, so that you may keep striving to understand its message, and keep working with others to conserve it.

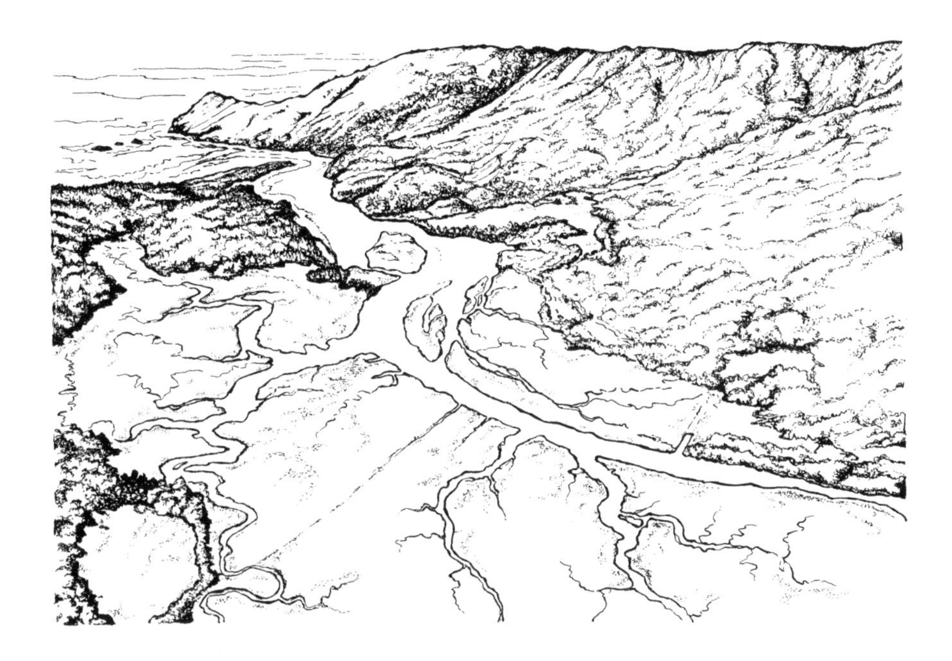

Chapter 9
What Is the Future?

You cannot step twice into the same river, for other waters and yet others go ever flowing on.

—Heraclitus (ca. 500 BC)

Shards of sunlight glint from the waters covering the broad tidal wetland of the Salmon River as I paddle my sea kayak steadily upriver on a brilliant October afternoon, an interlude between the rainy days of fall that settle in along the coast of Oregon. The freedom to glide unannounced over the shallows and glimpse nature undisturbed feeds my life-long love of hand-powered watercraft. The blades of tufted hairgrass and Lyngby's sedge that cover this marsh over which the tidewaters rise and fall are mixed gold and green now, in beautiful contrast with the backdrop of dark velvet Sitka spruce and western hemlock forest on the headlands beyond. Dark green-brown Dungeness crabs scavenge across the bottom only a few feet below, as I roll upstream on the incoming tide. After a few miles I rest a bit, smell the faint salt in the air, and ponder the river, middle-aged arms and shoulders aching from the work. Suddenly, my reverie is shattered as Double-crested Cormorants, looking almost reptile-like with their scaly jet-black feathers, leap from their small committees on the floating snags and beat the water with each downward stroke of their powerful wings as they take off low across the water. Curious harbor seals follow me miles upriver as I paddle, and my throat catches in surprise when one pokes its head above the water just ahead to inspect my boat. No such crabs or seals ply my home rivers.

But the main event in Northwest rivers like the Salmon at this time of year is the upriver migration of spawning adult Chinook and coho salmon, two of the five species of Pacific salmon in North America. Chinook, the behemoths among salmon, reach forty pounds in this river and are prized by anglers such as those gathered near the highway bridge as I journey upriver. Just the week before, I was privileged again to witness something primeval when I happened upon a large female digging her nest in a shallow riffle along the headwaters of another coastal stream nearby. I am always surprised to meet such large fish in small streams. How long would the white cartilage rays in her frayed tail last as she turned on her side to beat and lift the gravel to winnow away the silt? I crept behind a clump of grasses to avoid disturbing her while she gradually, patiently, dug a shallow nest (called a "redd") in which to bury her pea-sized salmon-pink eggs with clean gravel so they could incubate safely through the winter. How would spawning salmon like her have avoided the brown bears that for millennia must have fished along these streams? The males, skulking nearby, somehow knew just when to slip by her side to release their sperm as she released her eggs. Then the female gave the last of the ocean's energy she had stored in her body to flail the gravel upstream of the redd and bury her final batch of eggs, before drifting away to die, as do all Pacific salmon after they spawn.

The fact that these wild salmon are still here, and move through the estuary and far upstream into the headwaters to spawn every year, represents a triumph for conservation of Northwest rivers like these. In the early 1960s, a square-mile area of salt marsh habitat in the Salmon River estuary was turned into pasture for cattle, about two-thirds of the tidal wetland habitat originally present. A network of earthen dikes was built to prevent salty water from flooding the marsh on rising tides, and one-way tide gates were installed on the mouths of all the small streams that descend from the headlands and wander across the estuary. At the same time, wild adult salmon entering coastal rivers like the Salmon gradually declined over the last fifty years, for a host of reasons that we know better now than we did then, including high rates of commercial fishing and destruction of habitat critical to young salmon in headwater streams and estuaries like this one. A hatchery was built on the Salmon River in the 1970s and released juvenile salmon to boost production, so thirty years later more than half the adult Chinook and about 90 percent of the adult coho salmon that returned from the ocean were hatchery fish.

But a broad-based group including local citizens, biologists with state and federal fisheries and forestry agencies, and legislators had a vision of a different future for the Salmon River and its salmon. Together they worked to prevent unwise development along the estuary, bought the land, and removed dikes and tide gates from the pastures to let the tidewaters return. A small amusement park built on drained lands at one edge of the marsh was removed and this land also returned to wetland. This painstaking work, which began in the late 1970s, has now restored more than two-thirds of the lost marsh area to the tidal flows that sustain it. Hatcheries can also reduce survival of wild salmon stocks and cause declines, so fisheries biologists decided to stop producing and releasing hatchery-reared coho salmon. Starting in 2009, only wild coho returned to the river, and biologists are working to rebuild the original populations. The timing couldn't be better, because in early 2008 coho salmon along the Oregon coast were listed as threatened under the Endangered Species Act.

But what is the connection between wild salmon and the estuary? Don't salmon spawn in freshwater streams, and their young later go to the ocean to grow into the large adults that return? Why would they need the estuary? Years of scientific detective work by fisheries biologists Dan Bottom (NOAA Fisheries) and Kim Jones (Oregon Department of Fish and Wildlife) showed that not all young Chinook and coho salmon follow the same career path. They tagged thousands of young salmon in the field with tiny electronic tags to find out where they go. They also removed the tiny inner ear bones (otoliths) from adult salmon after they spawned to analyze the chemical signatures that were laid down from the surrounding water as the salmon grew. Each type of aquatic habitat has a different concentration of trace chemical elements like strontium, and these differences can be measured in the layers of the otoliths using lasers and sophisticated laboratory instruments. From this chemical analysis, they were able to discover at what age the salmon migrated downstream as juveniles from freshwater streams to the estuary and into the ocean. When they combined this with records of the tagged juvenile salmon that passed over underwater antennas as they migrated downstream to the estuary, and those they recaptured in the river and estuary using traps or nets, Bottom and Jones assembled an amazing picture of just how different the "life histories" of the young salmon can be.

Based on early research in other rivers, fisheries biologists assumed that most juvenile coho salmon and many juvenile Chinook feed and grow in the freshwater streams where they were born for a few months to a year, and then emigrate downstream rapidly into the ocean, using the estuary at most only briefly. Although Jones and Bottom reported that many young salmon do follow these typical life-history patterns, they also found that the salt marsh and the tidal channels that run through it are critical for the growth and survival of other groups of juvenile salmon during their first months of life. In fact, they discovered that both species of salmon had a total of four different life-history types of juveniles, each using different habitats at different times and migrating to the estuary during different seasons. These ranged from salmon fry that emigrated downstream soon after emerging from the gravel redds and reared in the estuary for extended periods, to others that lived in headwater streams for a full year before migrating rather quickly through the estuary and to the ocean as yearlings.

Among the juvenile coho and Chinook salmon that rear in the estuary, Jones and Bottom discovered that several of the different life-history types of each species move downstream during their first spring or summer and live in the complex of anastomosing channels draining the estuary for a few weeks to a year. They gorge on the rich array of aquatic invertebrates produced there and grow rapidly as they prepare for life in the vast ocean. For example, one of the life-history types of coho salmon migrates downstream to live in the estuary during their first summer, but then moves sideways through the salt marsh and into the tiny headwaters of freshwater streams to spend the winter, before finally moving back downstream to the ocean in the spring as yearlings. Overall, the records from the otoliths of adult coho salmon that return to the Salmon River to spawn show that up to a third used this estuary habitat for substantial periods when they were juveniles. This complex array of different life histories allows these two species of salmon to fully exploit all the food and habitat resources of this relatively small coastal Oregon watershed to sustain their populations. In effect, it is like having eight different species of salmon coexisting in one river.

But why are these different life-history types important, not only to the salmon but to us as humans? Why not just make large numbers of one type in a hatchery and release them to the ocean to produce lots of adult salmon? For the salmon, the reason is the same as for all things human—our plans, our money (investments), and indeed our lives. It's never a good idea to put all your eggs, salmon or otherwise, in one basket. When a tsunami buries the tidal marsh with several feet of sediment in one year, as it has every few hundred years over the course of millennia, the juvenile salmon that rear in headwater streams provide insurance against such a disaster. In contrast, the types of juvenile salmon that move to the estuary soon after emerging as fry may help sustain the population when a severe summer drought dries most of the habitat in headwater tributaries. And with climate change looming, we humans need all the diversity in salmon life-history types that is possible, to provide resilience against the extreme changes that are now occurring and predicted for the future. Other salmon biologists have likened the advantage of conserving these many types to maintaining a diverse portfolio of stocks, bonds, and other investments to buffer against downturns in global financial markets. Just as diverse investments are a key to a secure financial future, a key to maintaining resilient salmon populations is to restore and conserve the freshwater and estuarine habitat needed to sustain these diverse life-history types. This may be our best hope to ensure that our children's grandchildren can witness their spawning in the Salmon River headwaters.

∽

Throughout this book I attempt to use stories and images to draw you, the reader, across the reflective boundary into the world of streams and rivers. I want you to meet some of the scientists who have studied and attempted to conserve them, from Shigeru Nakano in Japan to the many scientists described in this and other chapters, and show you what they have learned about how streams work. But I also want to show you what we as aquatic ecologists bear witness to daily—that many streams, and the organisms that live in them, are declining rapidly and predicted to decline even more. They are already in much worse condition than many of the terrestrial ecosystems we see around us every day. And in the end, I argue that the streams we are losing are critically important for us as humans if we ourselves are to be resilient and whole. Rivers not only provision our water and food, but also fulfill an evolutionary legacy of unspoken need, and feed and heal the deepest longings of our souls. So, if rivers are this crucial for so much of what sustains us, then only one question seems at all important to me, cutting through to the heart of the matter. What is the future for rivers and streams? And if, as Leopold instructed, the important thing is to strive, then how can our striving more often be as successful as the restoration of estuarine habitat in the Salmon River?

Some people will argue that having real streams is just too difficult and expensive, and not realistic. After all, there are more than 7 billion humans and we all need water—for drinking, washing, growing crops, and supplying industry. Especially in dry climates like Colorado where I live, the streams and rivers we have are simply the difference between supply and demand. Of what nature supplies, our rivers are what remain after we humans have taken what we want and believe we need. Give this reality, some people may ask, "If you say these ecosystems are critical for us as humans, why can't we just create some new streams?" For example, if the $400 million of agriculture produced each year in the Arikaree River basin ultimately feeds many hungry people and supports an entire human economy in eastern Colorado, then wouldn't it be cheaper and easier to just pump water out of the ground somewhere else and create a river for the fish and wildlife, and for us to look at and listen to?

I'm quite certain that we won't be able to create any new rivers. At least, we won't be able to make any that will sustain the multitude of beautiful living organisms we expect to find there, from brassy minnows to ground beetles, and willows to warblers. Rivers and streams and their riparian zones are fundamentally a product of the geology underneath the ground and the way that water moves over the surface of the land and beneath it. In short, they are produced by the hydrology of the basin that sustains them, and this hydrology cannot be reproduced in places other than where rivers already run, at least not without human technology that risks going awry. And it is this complex physical system of seeping and flowing water, interacting with the landscape, that provides the conditions necessary to support the plants and animals on which the rest of the food web depends. Without it, the food web will collapse.

But it's not as if people haven't tried to create new rivers, and argued that they were successful. In the southern Appalachian Mountains, entire mountaintops are torn down to get at layers of coal beneath, and all the rock removed is dumped into nearby valleys. The forested valleys in these ancient mountains once sheltered small graceful streams and their riparian zones, and the long span of evolutionary time produced the highest numbers of fish, mussel, and salamander species of any region in North America. A recent federal rule allows digging ditches on top of this valley fill, called "stream creation," as valid "mitigation" for burying more than 1,200 miles of streams in the central Appalachians by this mountaintop mining. But the water that collects in these artificial channels when snows melt or after rains just sinks out of sight into the valley fill. And after this groundwater percolates through the newly exposed waste rock and eventually seeps out somewhere downslope and downstream, it is often toxic to aquatic invertebrates and fish. Even if the ditches held water and the water were not toxic, the channels are not shaped by floods nor bordered by native riparian vegetation. After all, the riparian vegetation itself requires the right kind of wet soils to germinate and grow. The upshot is that these artificially created "streams" support none of the habitat, biota, or ecosystem functions of the streams that the mining buried deep underground. No one would consider going fishing or bird watching there, having a picnic, or walking along them to find solace after a grave loss.

In the West, we move water around to create and alter rivers. Engineering schemes move water from places where it is plentiful to cities and farms where it is needed, to support humans in this dry climate beyond the 100th meridian. Could such schemes also be used to create the rivers that we want and need? For example, if we want the Arikaree River to flow along its length again, can the water pumped from the ground to grow crops simply be replaced by diverting water from the South Platte River, the next basin to the north? Unlike the Arikaree, the South Platte headwaters begin high in the Rocky Mountains, and melting snows supply this river with huge amounts of water that flood east onto the Great Plains in early summer. Couldn't more of these floodwaters be stored in upstream reservoirs, and then later released and diverted across the low divide to supply stream flow in the Arikaree River during late summer?

Unfortunately, as in most human endeavors, "The best laid schemes o' mice an' men, gang aft agley"—that is, often go astray. At least three problems can occur when water is moved from one basin to another like this to meet our needs. First, and most obvious, is that we simply transfer the problem of altering flows somewhere else. Loss of early summer floods in the South Platte River caused by storing more floodwaters has already drastically changed riparian forests for hundreds of miles across the Great Plains. Originally, most tree seedlings were washed away by the annual floods or died when the riparian soils dried up in late summer. Early explorers like John C. Frémont found cottonwoods only few and far between, isolated in wetter places like hollows and ravines, when his party traveled up these river valleys in the early 1800s. But the more constant river flows created by agricultural irrigation have allowed tree seedlings to survive, many of them not native to the region. Now there is a continuous strip of riparian forest across the plains, which has allowed nonnative animals from the East such as the Blue Jay, Baltimore Oriole, and white-tailed deer to invade the Rocky Mountain region and hybridize with native species in the West. We also divert water from mountain headwater streams on the "western slope" of Colorado and move it through huge pipes tunneled beneath the mountains to supply those of us who live along the Front Range east of the Continental Divide. The current and new proposed "trans-basin" diversions are drying up the headwaters of the Colorado River, one of the premier rivers in Colorado for trout fishing and other recreation.

Second, moving water between basins brings the possibility of introducing nonnative fishes, invertebrates, plants, and diseases. For example, the nonnative zebra mussel, a tiny mollusk from Europe, was brought to the Great Lakes by oceangoing ships. However, it then also invaded the Mississippi River system through the Chicago Sanitary and Ship Canal that connects Lake Michigan to the Illinois River near Chicago, where zebra mussels were first sighted in 1991. Within only two years, it had dispersed 1,400 miles south, to the mouth of the Mississippi River in Louisiana, probably by attaching to commercial boats. The costs of damage by zebra mussels, which coat water intakes and power plants and alter food webs on which fish depend, are at least hundreds of millions of dollars per year for the Mississippi River basin alone, and $1 billion per year in the United States. Many fish invasions have also been caused directly or indirectly by structures that move water among basins.

A third problem is that humans often change their minds, especially when costs go up. Moving water to support river flow for fish or human "aesthetics" depends on technology, infrastructure, and energy for operating and maintaining dams, reservoirs, ditches, canals, and pumps. When technology fails, and the cost of energy rises, will we still be interested in providing water for fish, birds, and the need we humans have to see, hear, and touch rivers?

We ventured forth from our hotel into the sultry night air of Valencia, the third largest city in Spain, and found our way into the sunken riverbed that now harbors the Jardín del Turia (Garden of the Turia). Lovers courted as they strolled along the weaving paths, and families with children played around the beautiful fountains and rich flower gardens. My colleague and guide, Emili García-Berthou, a well-known fish biologist and professor from Girona University in northeast Spain, explained that the Turia River that once flowed here created massive floods several times during the last hundred years and many people had been lost. The most recent, in 1957, flooded the city 8 feet deep and killed at least eighty-one people. "To avoid this flooding, the entire river has been diverted south of the city," he said. From our vantage point we could see the buildings of the Ciudad de las Artes y las Ciencias (City of Arts and Sciences), amazing designs constructed by famous architects in the original riverbed, which house a science museum, an aquarium, and an opera house.

But as we walked, I also noticed something else. The landscape architects who designed the beautiful gardens had included a cement-lined artificial stream, meandering along the edge of the paths, complete with artificial plantings of reeds and cattails in submerged containers and riparian shrubs watered by drip irrigation. The water appeared sanitized and in it we noticed a few frogs. But no fish rose gracefully from the depths of pools to catch drifting invertebrates. No birds or bats darted along the banks to catch emerging insects, and no spiders wove webs in the shrubs along the stream banks. Pondering this scene, I wondered, why would humans spend money to create features like this, which are relatively common in urban parks, unless they consider streams essential in their landscapes? What of value has been gained, and what essential values lost, for reasons we are unable to fully describe in words? And, if it is clear that we are unable to construct streams that will sustain what is essential for us, then isn't conservation of the ones we have a better route?

There is something priceless about rivers that leads us to conservation. Again, Leopold lends a clear voice to this difficult linkage with his poetic prose. "There must be some force behind conservation, more universal than profit, less awkward than government, less ephemeral than sport, something that reaches into all times and places where men live on land, something that brackets everything from rivers to raindrops. . . . I can see only one such force: a respect for land as an organism . . . out of a sense of love for and obligation to that great biota. . . ." When we reduce water to dollars per acre-foot, we reduce this conservation to a trivial meaning. Even if we could sufficiently re-create many of the ecosystem services supplied by rivers (which to date we have been unable to do), and therefore could put a price on them, are we ready to trade others that we cannot even describe away for mere cash?

The things that give our lives value often involve great personal sacrifice. Philosopher Viktor Frankl argues that it is this sacrifice, such as to nurture and protect one's family even in times of great peril, that gives one's life meaning. Would I sell my children or my spouse whom I love so dearly, or give away my most important photographs? In like measure, could I ever be happy with the prospect of no more rivers in the foothills near my home, as the water is sold and diverted away? Each of these things gives my life meaning, and these are things one would gladly make great personal sacrifices to sustain. "Our moral responsibility to care for the land grows from our love for the land," writes Kathleen Dean Moore in her essay on the ecological ethic of care, and it is no different for rivers. And throughout her writing, she then asks the next moral question of us as readers. So, given that we love the land, and its rivers, how should we act?

What are the right questions to ask, the best framework to hold, the most effective courses of action to take? In my view, the right question to ask is "How can we have real rivers and streams in our landscapes?" Although a detailed treatise on the theory and practice of river restoration and management is beyond the scope of this book, a more important goal is to seek a basic conceptual framework to guide thinking, speaking, and action (including voting). To that end, I present here four key strategies for conserving real stream-riparian ecosystems, based on a mixture of common sense and many decades of research and practice by stream ecologists and managers.

As a first strategy, our goal could become to not waste our future. Indeed, much of the water we divert, energy we use, crops we grow, and the land we mine to provide the resources we want is often simply wasted. By one estimate, 4 million pounds of material are mined, extracted, wasted, and disposed of to provide the average American middle-class family with what it uses each year. Essayist, philosopher, and Kentucky farmer Wendell Berry writes of our soil resources, "To waste the soil is to cause hunger, as direct an aggression as an armed attack; it is an act of violence against the future of the human race." I feel the same about water.

And, in addition to the direct uses of water for drinking and to grow crops, more than half of the freshwater withdrawn from rivers and lakes each year (53 percent) is used in generating thermoelectric power, as cooling water for the steam turbines. The hot wastewater produced is then itself cooled in huge cooling towers and either recirculated and reused or released back into a river or lake. However, large volumes also evaporate from the towers—8,000 gallons per minute from one large Georgia coal-fired thermoelectric power plant. In addition, by far most water used to make electricity is not withdrawn from rivers but is run through hydroelectric turbines built into dams to generate power. So, wasting electricity is wasting water, and wasting habitat in free-flowing rivers by damming them.

The good news is that 81 percent of the water that is *consumed* (that is, eventually evaporated and not returned to rivers or streams or lakes) is used to irrigate our crops. The reason that this is good news is that irrigation can be made to be much more efficient. If we are willing to pay the cost, water can be delivered in enclosed pipes to drip irrigation systems that water individual crop plants or trees, instead of in water-wasting systems like open ditches from which water is siphoned into long rows of corn or sprinklers that spray water that evaporates before reaching plants. Many farmers are already finding innovative ways to save water.

Likewise, huge amounts of water in rivers could be saved from being diverted to cities in arid regions by citizens who simply do not water their lawns every day during the growing season. I was surprised to find that during a recent drought, my city met its goals for water conservation by limiting lawn watering to every other day. I had rarely found it necessary to water more frequently than this, even to keep the one patch of bluegrass in my yard a bright green. In a recent plan for building more reservoirs and diverting more water from our local river, the proponents assumed that each person needs about 210 gallons of water per day. In contrast, actual water use per person in southeastern Australia (New South Wales, the state with the highest water use) is 52 gallons per day, only a fourth as much. In Spain, another dry climate, use is similar to Australia at about 41 gallons per person per day. It is well known that even those Americans living in wetter climates use two to three times more water than people in other developed countries with similar climates. Yet, people in those other countries seem to have sufficient water to sustain their lives and livelihoods.

To be happy, perhaps humans need streams more than they need showers that spray water from all directions. Perhaps we need the views and sounds of real rivers more than we need shopping malls and parking lots that are lit all night. Perhaps we need waterfalls to hear, fish to see and hold, and riparian zones flush with turkeys and bats and deer more than we need to grow more corn, much of which apparently becomes ethanol to fuel vehicles like oversized pickup trucks driven by we city dwellers. These are the real trade-offs we all face daily.

A second key strategy, in order to have real streams, is to afford them more protection than they now receive. Once protected, we need to find the most effective means of conserving our best streams and restoring damaged ones, and find a way to do this within a mix of environments that sustain human societies. These are nearly insurmountable tasks, but worth every effort we can expend to achieve them.

As it stands now, rivers and streams in many regions are vulnerable. For example, in Colorado, we humans have declared "water rights" that give us legal standing to dry up rivers entirely. These rights are granted in perpetuity, as long as the water is put to "beneficial use" to grow crops or sustain cities and industries. Because rainfall is insufficient to grow most crops in this arid region, the founding settlers realized that a system providing legal rights to the water available was essential to divide this precious resource among people and prevent physical fights and lawsuits. However, the main provision in Colorado, passed into law in 1879, was simply "first come, first served." Whoever could prove that they were the earliest to divert water from a stream and put it to beneficial use could claim the senior water right and get their water first every year. Nearly a century later, in 1973, a state law was passed that allowed protecting flow through a reach of river to preserve the natural environment. Cities and agencies could then file legal papers to reserve these "instream flow" water rights and provide habitat for aquatic and riparian organisms, something not allowed previously. Unfortunately, by that time so many water rights with more seniority already had been granted that most streams could be dried up at some time during the year, if everyone diverted all their water. Fish and riparian zones had come to the table too late. In the jargon of the profession, most streams were already "over-appropriated," and had been since the 1890s.

What is our true "water right" as humans? Do we have a legitimate right to use water for drinking and washing, growing the crops and animals we need to eat, and supplying the industries and regional economies that provide goods and services that are essential to sustain healthy human communities? This makes sense. I argue that we do have these rights. But do we also have the right to use water that ultimately goes to waste?

For example, in climates where irrigation is required, about 65,000 gallons of water are used to produce one of the big round bales of hay that lie in rows along hayfields in rural areas. Some of this hay is never used and rots in the field. More than 700 gallons of water are required to grow the cotton and process the cloth to create the T-shirt that I may wear only once and then give away to charity, perhaps never to be worn by anyone else. Do we have the right to water our lawns carelessly or to excess, so that water runs down the gutters in our neighborhoods? To waste water in ways like these is to deny our rights to have real rivers and the rich wonders and benefits they afford.

Protection of rivers is needed for more than just the water. Protection is also critical for the many linkages within streams and those between streams and their riparian zones. These are the threads in the ecosystem web that supply food and other resources to fish and salamanders and otters within the aquatic realm and to birds and bats, lizards and spiders, in the riparian beyond. And yet, careless or malicious introductions of nonnative bait, sport, or aquarium fishes by anglers and others can destroy these linkages. These invaders in aquatic systems are the equivalent of germ warfare, because those that invade can reproduce rapidly and spread over long distances throughout drainage basins. Many states in the United States issue a relatively small fine, equal to only a slap on the wrist, for such introductions, even though they can disrupt river and lake food webs forever. Such penalties are often far less than those meted out for illegally harvesting one trophy deer or elk when it is not the hunting season, which has few long-term ecological consequences. Likewise, we are only beginning to consider the potential effects of introducing large numbers of hatchery-reared fish on river food webs. Doubling the density of fish doubles the number of mouths to feed, and predation by these added fish may cause trophic cascades and other indirect effects throughout aquatic and terrestrial food webs, similar to those we found during our research in Japan.

Unfortunately, rivers are not well protected from most of these human actions, ranging from overuse of water to create products that are ultimately wasted to inadequate education, policies, and enforcement that lead to introduction and invasion by nonnative species. Myriad other activities can also create major problems, such as use of toxic chemicals on lawns and crops that ultimately drain into waterways and contaminate the aquatic food web. Hydropower dams change flow regimes, which in turn degrades habitat and alters the life cycles of fish and invertebrates that are cued by these flows. Overall, many current human actions are directly or indirectly at odds with sustaining the form and function of these ecosystems on which we depend for our health and well-being.

But, if we can manage to keep sufficient water in stream channels and keep nonnative species and pollutants out, how should we then manage the landscapes in these watersheds to have real streams? After all, many of our most important uses of lands, such as for agriculture, forestry, mining, and urban areas, can have profound impacts on streams and their riparian zones. For example, streets and parking lots of urban areas prevent rain from percolating into the groundwater and supplying streams with clean and constant flow. The water that drains from these hard surfaces during storms can create flash floods that make gullies of streams if areas are not set aside to collect this water and let it settle gradually into the ground. Long ago, Eugene Odum, one of our most famous ecologists, proposed that humans need a mix of environments to sustain societies, including not only protected places, but also productive ones for agriculture, forestry, and fisheries, and urban-industrial ones and compromise environments. The trick, as Odum described, is to achieve the optimum balance among these uses. Because the landscapes that support our riverscapes also support humans in most cases, it is clear that we must think of landscapes as integrated ecological and socioeconomic systems. Sustaining resilient ecosystems that can support rivers can only be accomplished if we develop human organizations that can learn, adapt, and create effective leadership to move toward this goal.

A main question that arises from Odum's scheme is whether it is better to attempt some balance between conservation and human land uses in "working landscapes" or to protect a subset of undisturbed streams while using others intensively. For example, if our wilderness areas and national parks were large enough, could we protect most of the organisms and values we seek in rivers? Or, do we also need to protect the many streams and rivers that traverse our productive, urban, and compromise environments? It is true that we have learned a lot about protecting streams in watersheds that we use for farming and forestry, primarily by creating buffer strips along streams. These strips of native vegetation, ideally at least 100 feet wide on each side of the stream, can trap sediment and pollutants eroding from fields and woodlots and prevent them from reaching streams. The riparian vegetation that grows there also shades the stream and provides the large woody debris and leaves that help sustain stream habitats and food webs. However, it is also true that larger streams, which collect water from many smaller tributaries, are increasingly difficult to protect from human abuses upslope and upstream.

In fact, this is a source of wonder and yet the crux of the problem with streams—they collect water and materials from many places far beyond and unseen. For example, the Cache la Poudre River that runs through my city collects water from about 150 small tributaries in the watershed, each a few miles long, and many channels of intermediate size. Erosion or pollution along only a fraction of these could have strong effects on my home river and its links with the riparian forest. In addition, many of the great diversity of aquatic and riparian species in any basin live in and along the larger, warmer, and more productive lowland streams and rivers, not in the small cold headwater streams that we often protect in national parks and forests. So, if we cannot protect more of the rivers and streams near where we live and work, then we cannot protect most of the animals that we value and that are within easy reach of our homes.

In the end, there is no simple answer to this dilemma of what to protect and how. If we want to sustain the full range of plants and animals associated with rivers, this often requires relatively undisturbed habitats. The reason is very simple. Some species in all flowing waters, from tiny streams to the largest rivers, are very sensitive to the problems that we humans create, such as increased sediment and changes in the flow regime. It is, therefore, worth our every effort to conserve the best places we have. Like human health, once it is lost, it is difficult to get all of it back.

Unfortunately, in many regions of the world, including many places in North America, there are very few undisturbed streams to conserve. Throughout most urban areas and vast expanses of agricultural lands, most streams are highly altered or degraded. And yet, these streams represent key opportunities to create what Odum viewed as compromise environments, by restoring habitat that can support more biota, supply cleaner water, and provide more aesthetic benefits than they do now. We should never let what we cannot do prevent us from doing what we can do.

Each time I travel to Japan, I marvel at the rivers that traverse urban centers like Osaka and Kyoto, where millions of people live and work in high density. At the end of one busy trip I boarded the ultramodern train to Kansai Airport that serves the region, which is situated on an artificial island created just off the sea coast to conserve land. For the last half hour of the train trip south, I descended ever deeper into the dense urban grid. I left behind views of lawns and gardens and forested hillsides and was met with only streets lined with offices and shops. Soon, I looked out on a densely packed jumble of traditional Japanese homes with tile roofs and modern office buildings of many floors.

I imagined myself as a child growing up in this city and wondered where I might have found trees and nature to feed my innate curiosity. As we neared our destination and gradually slowed, suddenly a river came into view as we crossed, with a narrow rich green riparian strip to contrast with the tan and gray of cement and steel. It was bordered by streets on both sides, as are most rivers in most cities, and storm drains poured in from both sides. But standing in the lush grasses at the edge of the channel in plain sight was an elegant white heron, stock still on jet-black legs, waiting for a small fish to stray near enough to strike. That image is burned into my memory still today.

It is worth much to humans for cities to provide access to their rivers, even as they work to restore them. In many cities in Japan and throughout the world, rivers and streams are encased in deep cement troughs and are impossible or dangerous for humans to reach. In part, this is because storm water that has nowhere else to go can create flash floods in these cement canyons and catch visitors unaware. But where storm water is properly detained and dispersed, and rivers are allowed a semblance of a floodplain over which their floodwaters can spread out, such as along the Kamo River in Kyoto City, the use by people can be very high, and highly rewarding. In late July a few years ago, several Japanese and American colleagues and I found many children and families wading and snorkeling in the cool, clear water around the famous large stepping stones that span the river, watching and netting small fish while cooling off from the sweltering summer heat. What child living in an urban area could have asked for more?

Cities are better places to live for having restored streams and rivers running through them, even if they can never be pristine. In the research described above by Spanish scientists who study sound, simply adding the sound of running water from a stream caused people to rate urban scenes more highly. These scientists proposed that the sounds caused people to perceive other values when watching these scenes, specifically the qualities of survival and fertility. Likewise, Swiss researchers found that people in that country preferred sections of rivers that had been restored, over those not restored, when they watched videos that included both sight and sound. The reason was apparently that restoration created complex and interesting habitats that were attractive both visually and acoustically, especially at a range of medium flows. These complex habitats and moderate flows with their appealing sights and sounds are among the first things lost when rivers are altered. Habitats become simplified, and flows are either very low or, infrequently, very high when the river floods. People in cities need rivers restored, and to hear and feel as well as see them, to gain the most benefit of what they offer to us as humans.

Will restoration of degraded rivers be easy? I can guarantee it will not be. Stream ecologists Emily Bernhardt and Margaret Palmer and their colleagues analyzed basic data from tens of thousands of stream restoration projects throughout the United States and found that most were never monitored adequately (unlike the work described in chapter 3) to find out what worked and what didn't. Companies and agencies that attempted to restore only the *structure* of rivers by modifying the channel to look more natural often ignored the *functions* that go missing because of the hidden connections or processes that are not restored. Without these critical functions, streams and rivers cannot support the aquatic and riparian biota that we seek when we visit them. Effective restoration, and monitoring of the outcomes, requires well-trained geomorphologists, stream ecologists, and economists and other social scientists working together. If we don't have the time and money to do the job right, do we have more time and money to do it over?

If restoration is difficult to do right, will it be worth the effort and cost? Even many professional biologists seem to think that stream restoration is simply "feel good" work, which is not effective at restoring habitat or recovering the fish and wildlife that were lost. Although the work is challenging, I believe it can be successful, and careful studies have shown this. Science, properly done, *can* provide answers about which restoration strategies and methods restore both structure and function, but only when the science includes all the important disciplines and considers carefully the mix of processes that create habitat in a specific riverscape. Because these processes are different in different regions, one size will not fit all. Much of the best is yet to come in this field of stream restoration. Yet even somewhat imperfect restoration focuses attention on local streams, garners public support to prevent further loss of flow and habitat degradation, and generates a community of citizens interested in creating resilient riverscapes in which their children and they can take pride and find solace.

‹ ›

A third strategy needed to ensure real rivers in our landscapes, beyond conserving and restoring stream flow and healthy habitat, will be to maintain key processes that sustain the functions and foster resilience in these ecosystems. But like ensuring flow and protecting habitat, this task will be daunting in many places. The simple reason here is that these key processes include floods, landslides, fires, wind and ice storms, and, yes, even droughts. And although ecologists call these processes "natural disturbance events," most people consider them to be highly destructive, even in wilderness areas. Most people view the aftermath of these events in streams as ugly, and think mainly of the years that it will take for the landscape to recover, but ecologists have discovered that these natural disturbances are critical in sustaining the organisms that have evolved to cope with them.

About thirty-five years ago, it became clear to ecologists that these natural disturbances, although destructive in the short term, also open up new spaces and provide new resources that allow less competitive species to survive in landscapes and riverscapes. For example, without windthrow that knocks down mature trees in old-growth forests, other species of trees and smaller plants that need light gaps would die out forever, shaded by the more dominant species that make up the canopy. Likewise, without floods to disturb river beds and wash the silt from gravels, not only does stream habitat deteriorate but dominant species of invertebrates take over, crowding out others that are important members of food webs that feed fish.

For example, the four-level food web that Mary Power and her colleagues studied in the Eel River of California (see chapter 3) persisted only when massive natural winter floods "reset" the system by churning the gravel and cobble streambed. The floods in this Mediterranean climate roll and grind the rocks against one another, a natural disturbance that kills many large stream invertebrates and temporarily scours the streambed free of algae. However, the floods also supply nutrients to grow new succulent algae and provide space and food to support smaller fast-growing mayflies on which trout depend. In contrast, in years when winter floods were small, Power and her colleagues found that a large grazing caddisfly that builds a heavy protective case of tiny stones took over, dominated the rock surfaces, ate most of the algae, and crowded the mayflies out. Unfortunately, trout are not able to eat or digest this large caddisfly because of its heavy case, so most of the energy flowing up through the food web in these years simply reached a dead end in this invertebrate. Other studies by Mary and her colleagues also showed that on other coastal California rivers where winter floods are controlled by dams, large heavy-cased invertebrates that fish can't eat dominated the food web every year. As a result, controlling the natural flood disturbances with dams can cause a collapse of the food web that supports fish like steelhead trout, the ocean-going form of rainbow trout which is listed as threatened or endangered throughout California under the Endangered Species Act.

Perhaps placing disturbances in a more familiar context would help us understand why they are needed. After all, on land we often disturb the soil by tilling it to increase the productivity of crops. Tilling suppresses weeds, enhances soil structure, and releases nutrients that plants need. We also know that prairie fire, another natural disturbance, gets rid of suffocating dead plant litter and releases nutrients into the soil, causing native grasses to flourish and wildflowers to bloom profusely afterward. In rivers and streams, in addition to the benefits of floods described above, other disturbances such as natural landslides bring whole trees and large amounts of gravel and cobble into streams, which are critical elements needed to create habitat for invertebrates and fish. Natural disturbances by wildfires in watersheds can also cause large pulses of sand and gravel to enter streams, because burning up the vegetative ground cover promotes erosion. This input can be important for streams in the long term, even though they may be inundated by too much sediment in the short term. Subsequent floods wash the finer sand downstream and rearrange the coarser gravels and cobbles to create the pools, runs, and riffles that are needed by invertebrates and fish.

Humans often attempt to circumvent these natural disturbances, usually because they destroy property and infrastructure built near streams such as homes, farms, bridges, and roads. For example, water from peak flows can be diverted through floodways or stored to supply water for use by cities, as has been proposed for my local river. Such schemes may seem appealing and are sometimes necessary to avoid damage by floods. However, they can also lead to degraded habitats, because without floods, pools and the gravel in riffles are filled in by fine silt and sand. Likewise, without floods that create clean open sandbars along the edges of western rivers, seeds of native cottonwood cannot sprout. When this happens, these beautiful trees that create riparian gallery forests eventually die out and are replaced by nonnative trees. And without floods to erode banks and fell large trees into rivers in all regions, fish and invertebrates lack the complex habitat and shelter from flow that these logs and their roots provide. Natural disturbances like floods, fires, and landslides are important for maintaining the character and function of natural rivers.

Are there success stories about using natural disturbances to improve stream ecosystems that humans have damaged? Can we restore natural floods, or mimic landslides and the sediment and whole trees they supply to rivers? Will this improve degraded habitat and sustain rivers and their aquatic and riparian biota? As in all of ecology and restoration, it depends. River managers and ecologists have released large pulses of flow to mimic natural floods, or modified flow in other ways toward more natural regimes, in about forty rivers and streams worldwide. Positive trends are noted in a variety of cases, but the results depend on how closely the restored conditions approximate the natural ones and whether organisms are able to recolonize. For example, flood pulses in several rivers worldwide failed to cause fish to migrate upstream and spawn as expected because water was released from the bottom of the reservoir, which was too cold. Fish are cold-blooded organisms, and their maturation and spawning behavior are highly dependent on temperature. In another case, the assemblage of invertebrate species in a Swiss river changed only gradually toward those better adapted for floods, even after a decade of flood pulses (twenty-two floods over a ten-year period). It turns out that the invertebrates lost from the river segment during the previous thirty years when flows were highly regulated were not able to recolonize easily by drifting or flying because the river segment was in a narrow canyon and isolated between two dams.

These examples show two things. First, river scientists and managers will need to use everything they know about these ecosystems and their biota to plan effective restoration using natural disturbances like floods. For example, often they will need to combine flood pulses with releasing water of adequate temperatures to allow reproduction by fish and invertebrates, removing levees that prevented floods from reaching productive floodplains and controlling nonnative fish that hamper native ones. Fixing only one problem will not be sufficient. And second, it is clear that as scientists and managers collaborate to get the details right, the future will be brighter than the past for many rivers where flows are being restored.

Streams and rivers change constantly, driven by processes that occur over large spatial scales and long time periods, so our habitat restoration must be planned and carried out in concert with this backdrop of change. In Pacific Northwest rivers of Oregon and Washington, for example, river managers are restoring habitat by mimicking the process of landslides that bring whole trees and tons of rock and soil into stream channels. In the natural condition, rivers then move these trees to stable positions, often lodged crosswise in the channel, where they collect thick "wedges" of coarse gravel and create deep pools. The gravel is perfect habitat for adult salmon to spawn in, and the pools provide excellent summer and winter rearing habitat for the juvenile salmon they produce.

To mimic these processes, river managers use large helicopters to bring whole large trees and place them in the river in clumps, at the base of gullies or tributary mouths where landslides naturally deposit debris. However, they then let floods do the work of moving and positioning the wood. Stream reaches that previously had no wood to create pools and virtually no gravel for salmon to spawn in now support many young coho salmon and steelhead every year. Not every log produces ideal habitat right away, but as they are moved downstream year by year and lodge crosswise or catch in log jams in new locations, the dynamic stream ecosystem creates and maintains suitable habitats in various patches throughout the riverscape. It is interesting that the Greek philosopher Heraclitus had already discovered these principles of river restoration about 500 BC, based on his most famous fragments of wisdom that have survived. Not only are rivers ever changing as they flow in a timeless continuum, as the epigraph for this chapter tells us, but tension and dynamics among opposing forces are also essential to create a kind of harmony in nature, like the two opposing arms of a bow working in harmony to propel an arrow.

( )

Finally, the fourth key strategy for conserving real streams and rivers, and the most important, is about us as humans. Rivers are protected from human excesses, and conserved for future care and use, not by the rivers themselves, but by people—by us. The Lebanese poet Kahlil Gibran, in his beautiful poem "On Children," used the metaphor that we are the bows from which our children are sent forth as living arrows, their souls to "dwell in the house of tomorrow, which you cannot visit, not even in your dreams." If any of our rivers are to join our descendants in their future, people will need to understand their dynamic nature, value them, and find effective ways to live near them without destroying them. Most of us take rivers for granted, until they are gone. We cross them daily on bridges, never giving thought to where they have come from and where they go, the water they provide us, or the life they support, including our own lives. We are surprised when they flood, even though this is one of the key processes in rivers. We are even more surprised when they run dry, especially if we depend on them to supply our water needs.

As I have written above, many authors have penned the thought that we humans conserve only what we love. Leopold's writings seem to be the source of these ideas for those who followed, including Stephen Jay Gould and David Orr. Leopold also argued that we can grow to love only what we have an intense curiosity to understand, the second point in the triumvirate later given us by the Senegalese environmentalist Baba Dioum. Standing at the threshold in time when ecology was becoming a modern science, Leopold strove to help his readers understand that the landscape supports a rich community of organisms within the broader ecosystem. He described this ecosystem as resembling a "round river," where elements that originate in the soil make their way through plants and animals and later, when they die, are returned to the soil and, eventually again, to "that great biota." The term "ecosystem," and the concept it embodies, were brand new ideas in the field of ecology in the 1930s when he first used them in his writing. If only he could help others understand the critical importance of these ideas for managing lands and rivers. Leopold saw this as the principal goal and need for conservation education, and Baba Dioum completed the message by translating this thought into words: "We understand only what we are taught."

But more is required of us than loving, understanding, and teaching about rivers. I am convinced that we cannot have real rivers until the public is speaking and writing directly to policy makers and land and water managers. Although the science that we stream ecologists have discovered, and the rich expertise embodied by professional aquatic ecologists who work for government agencies and nonprofit organizations, are critical elements in effective management, this expertise is often overlooked. For example, I can count relatively few times when those who turn the dials and make the decisions on water use and river restoration in my region have asked for advice on how our rivers should be managed to sustain real functioning ecosystems (although more advice is being sought now). The reason is that these decision-makers are busy, and respond to the loud voices and media attention that make up our current democracy, and always have.

In the mid-nineteenth century, Frenchman Alexis de Tocqueville wrote that when you first arrive in America, "you find yourself in the midst of a sort of tumult; a confused clamor is raised on all sides; a thousand voices come to your ear at the same time, each of them expressing some social needs. [. . .] Citizens assemble with the sole goal of declaring that they disapprove of the course of government.... To meddle in the government of society and to speak about it is the greatest business...that an American knows." Environmental writer and activist Bill McKibben has noted that more attention can be gained for an issue when a few hundred or a thousand people assemble on the steps of local governments, or join hands and surround the White House in peaceful demonstration, or organize local rallies via the Internet, than by many other forms of free speech in America. Politicians respond to their vocal and visible constituencies.

We will begin to achieve standing for real rivers when our voices rise in this visible clamor.

In the end, I am convinced that exciting our current generation to cross the boundary and understand the hidden mysteries that lie beneath the surfaces of streams and rivers, and educating them about their worth, will be the path we must pursue. My friend David O'Hara, who teaches philosophy, informed me that this is a very old idea, expressed elegantly by the rabbis who wrote the Talmud (here translated loosely). "It is not your job to finish the work, but neither are you free to walk away from it" (Pirke Avot 2:21). Only when we inspire them to cross this reflective boundary can this generation prepare the next to raise their voices in support of conserving real rivers and streams in which their own future generations can find solace and wonder.

‹ ›

My sabbatical time had grown short. I had spent six weeks of the fall of that year on the coast of Oregon at the Sitka Center for Art and Ecology, living and breathing the ebb and flow of the Salmon River and its estuary and pondering the half-formed ideas that I sought for this book. It was a magical time filled with learning new ways of writing, gaining new friends, struggling with familiar and new demons, and participating in gathering and eating the fruits of the land and its rivers, and the ocean. My writing was interspersed with meals of wild salmon, oysters, and chanterelle mushrooms, and long hikes on the headland overlooking the river.

At the dimming of the day in early November, on the eve of the end of my stay, I was again drawn down the hill to Knight's Landing to listen and watch along the broad river estuary, now at low ebb tide. A light rain fell and a bitter wind blew across the valley, as yellow alder leaves blotched with brown decay streamed earthward beneath the lowering clouds. A Great Blue Heron struck its silent pose, fishing in the shallows along the near bank. A harbor seal's head made a "V" in the soft silky reflection of the sky as it journeyed upriver. Along with my love and respect that had grown for this river in the weeks that had passed, I felt a sense of mourning, of what had already been lost by our influence, and what we might have lost forever had not committed people cared enough to restore the hidden connections with the tidal marshes. The salmon were coming back again, to fulfill their ancient quest.

But while lost in this melancholy, I suddenly heard the caviling of Canada geese, faint in the distance. It was the goose music that Aldo Leopold so loved, sung by birds beyond my vision's reach in the failing light. Then, despite the cold wind, my heart thrilled as, just beneath the low clouds, a ragged V came straggling into view, headed downriver, the cacophony growing louder. Without warning, the flock whirled over the velvet forest on the far shore, circled back, and all the geese set their wings to land, as if by some silent pact. They settled, wings flapping, onto the tidal flat, discussing their plans to forage on the river's bounty.

As I turned to leave, I recalled Leopold's lament, about the loss of wild things he so loved and hoped to leave his children. "And when the dawn wind stirs through the ancient cottonwoods, and the gray light steals down from the hills over the old river sliding softly past its wide brown sandbars—what if there be no more goose music?" Surely this sense of loss is the best measure of the things we love. Climbing back up to my cabin at dusk, I pondered this same question, but for the river itself.

Epilogue

Nakano's Legacy

Small, neatly terraced rice paddies shoulder the stream as we wind along the narrow road from Kamioka toward the headwaters in the steep-sided valley. Everywhere the landscape is lush, bright green, in these mountains of Gifu Prefecture in central Japan, where Nakano grew up. The broad peaks bring the Appalachians to mind, and the deciduous oaks and maples and the humid climate add to the similarity. Each rill and run seeps water, collecting into tiny streams, and ultimately joining the Takahara River that drains the region. These streams that begin deep in the mountains also harbor native whitespotted charr and masu salmon, the fish that the chef in his parents' restaurant told Shigeru about when he was a boy. As we drive, I ponder this landscape that shaped my friend's life and work.

Yukinori Tokuda stopped his SUV at the end of the road next to a grassy riparian meadow and found his boots in the back. He reminds me so of Shigeru Nakano—an older and wiser and gentler Nakano that I had only begun to glimpse near the end of our days together. Tokuda-san was Nakano's best friend in college, the one with whom he first snorkeled in streams to study fish. Tokuda's dream had been to be an underwater photographer, but as life and happenstance would have it, he became the fisheries biologist who manages these streams for the local private fisherman's union (the main groups that manage fish in streams in Japan). They had both returned to Nakano's hometown after college, and both married young women from Kamioka. The first few years they worked for the fisherman's union to survey the fishes in these small streams throughout the Hida Mountains, because most had never been sampled. But in their spare time they also each pursued their main passions.

Tokuda-san headed off briskly up the trail toward the headwaters of Utsubo Dani (Utsubo is a local village, and Dani means stream), eventually leading us through tall forest and past a few cabins situated along the stream. Daisuke Kishi (Nakano's former student), now a fisheries research biologist with the prefectural government, and me, and photographer Dave Herasimtschuk kept up with Tokuda's fast pace. We had returned to Japan in July 2012 to reconnect with Nakano's family and colleagues, gather material for this book, and visit and film the places and people that were important in Nakano's life. In particular, I had asked to see one of the streams where he had done his first research in those years after college. We squished along the trickles that ran down the path, and as we hiked I drank in the sights and sounds of the rich forested landscape. Tokuda did not slow throughout the half-hour hike, still as fit as in his youth. Across a bridge he strode, and then, turning upstream, he hiked another 50 yards through the riparian zone and clambered down through the head-high *sasa* (bamboo) to the base of a large pool, waiting for the rest of us to catch up.

As I emerged through the whips of bamboo, the thought came racing—*This is the place!* There was the large pool resting at the base of a 10-foot waterfall over sheer bedrock, looking just like the detailed map in his paper that I had pored over many times. The pool was larger and deeper than those we had studied in Poroshiri Stream, about 20 feet wide and 35 feet long, and 6 feet deep in the middle. *This was the place!* This is where Nakano had collected the amazing data that he showed me in October 1988 in Sapporo when we first met. And the paper he later published caught the attention of many other scientists too, because it described how two completely different species set up and defend territories to divide up space and food resources. Even now, twenty years after the paper appeared, it is a key example of this "interspecific territoriality," and is cited in one of the most widely used textbooks in ecology. Even Nakano's early work has stood the test of time.

Tokuda and Kishi and I stood for a long time in the cool misty air at the edge of the pool, talking of many things in highly broken English. Tokuda described how Nakano and he worked every day through long stints of field research that summer of 1987, living in a cabin nearby the stream. And then the same thought dawned on both of us at the same time, each in our own language. They had done that study here exactly twenty-five years ago this summer, when both were twenty-five years old. Since then, time had flowed onward, marriages and anniversaries had happened, and their children were born and had grown and started college. And Nakano had passed away, twelve years ago now. I could see the wistful look in Tokuda's face, now sadder but wiser, appreciative of all of life's wonder and bounty and grace, despite the difficulties. Each of us who had known him, Tokuda and I and Kishi, could feel Shigeru standing with us there in that place that had meant so much.

Perhaps more than other types of scientists, field ecologists are connected not only to other people but also to places. By definition, they do their work in specific places, and the singular places where important work has been done are imbued with great meaning. Like those houses where we grew up and those places where we first met our best friends and life's loves, the places where we and others have done our most meaningful studies hold for us our connections to these colleagues and our past, connections forged in the striving to accomplish the work and the deep insights gained. They provide a reservoir of remembrances, and on that day at Utsubo Dani our thoughts linked us to the spirit of our friend and mentor who was gone from this life. That memory, standing near the endless flow of those running waters, will haunt me always.

Throughout our journey in Japan that summer, from the oppressive heat of Nagoya and Kyoto to the beautiful mountain streams of Gifu Prefecture and the southern island of Shikoku, and all along the shores of ancient Lake Biwa, I pondered Nakano's legacy. What did he leave us? What had I come to understand anew and see freshly? What had I and we done that was good, and what would not have been done had I never come to Japan and met Shigeru Nakano?

Just a few weeks before, while attending the national meeting of the Society for Freshwater Science, where most stream ecologists gather, the science that he helped shape was all around me. Whole sessions of scientific presentations at that meeting in recent years centered on the ideas that Nakano and Mary Power and others had pioneered about how streams and riparian forests and grasslands are connected through the movements of aquatic and terrestrial insects. New graduate students formed their doctoral dissertations around testing where and when and why these connections are important, from the effects of cattle grazing on terrestrial insects that fall into streams and feed trout to the role of emerging aquatic insects in moving toxic chemicals from streams to birds and spiders in the riparian zone. The study of "reciprocal subsidies" between these linked ecosystems is now one of the most exciting fields in stream ecology, with many unique discoveries, including by new young researchers in Japan.

Tributaries that feed a main stream can often change its character markedly, by enriching it with life-giving nutrients or adding colder water that makes it habitable by trout. So like this, a scientist's ultimate contribution is strongly influenced by the confluence of knowledge and inspiration and collaboration from his or her mentors and colleagues. For me, personally, Nakano changed the ways that I thought about and conducted science, as I also changed these ways for him. Whereas I apparently taught him the value of field experiments, he offered me a much broader and richer view of how to be creative in research. My approach tended to be focused and conservative and timid, but he taught me to be more observant, think deeply, consider much beyond just streams and fish, and to ignore the criticisms of others. But most of all, for all of us, he embodied what it means to seize the day—*Gambareyo na!* (You must do whatever it takes!)

Life offers each of us so many opportunities that we had not planned on. For me, the opportunity to explore deeply the rich culture and history and language of Japan enriched many other parts of my life, and those of my wife and children. And even the tragic loss of Shigeru Nakano at the peak of his life and career led to greater meaning for those of us who carried on after him. That experience brought me closer to all in his family, and I forged lifetime bonds with his students and colleagues, as we sought meaningful and productive ways to heal our grief and transcend this loss. I have experienced the quiet joy of mentoring his former students in many small ways, like Satoshi Kitano, Yoshi Taniguchi, Mikio Inoue, Yoichi Kawaguchi, Yo Miyake, Masashi Murakami, and Daisuke Kishi, whom we visited on our journey. All are now accomplished professors at universities or scientists in research institutes. Nakano's pride in them was deep and unmistakable. In turn, they and others, like his colleague Katsuki Nakai, now a scientist at the Lake Biwa Museum, have mentored my postdocs and students and colleagues. Graduate students in Japan who never met Nakano but were influenced by his work have become young professors, and developed collaborations with the best stream ecologists in the United States. In turn, many of these American scientists were influenced by our research when they were graduate students and became inspired to collaborate with Japanese scientists. The web across our scientific cultures now has many tangled threads and is unlikely to be broken.

And as we rode trains through the heat and humidity, hither and yon across central and southern Japan that July, along beautiful rivers and sea coasts, through mountains and major cities, I reflected often on how improbable this journey had been, as improbable as our scientific discoveries. "Had I never answered that letter to attend the Charr Symposium in Japan, or had Nakano never approached me there, I would not be on this journey now," I thought. How unlikely it had been that Yoshi came to my university, and that Nakano had decided to visit scientists in the United States on his last trip. What if I had closed the book on Japan after the accident, or if Jeremy Monroe had not taken the huge risk to start Freshwaters Illustrated and make the *RiverWebs* documentary film? Life itself is the most improbable combination of good fortune, sweet and bittersweet.

Hiroya Kawanabe and I climbed steadily up the long series of wide rock steps toward Kibune Shrine, a 1,600-year-old Shinto shrine near a small stream in the mountains about 10 miles north of Kyoto. Kawanabe-*sensei* had witnessed my journey with Nakano and his colleagues since our earliest beginnings. He had convened the Charr Symposium in Sapporo in 1988 and was the key leader, had guided us and helped us gain our funding, suffered Nakano's loss deeply, and encouraged our efforts afterward to honor his legacy. Indeed, he is among the most renowned ecologists in Japan, and the world. I am in great awe whenever we meet, and yet he has always treated me as a close younger colleague and friend.

As we ascended to the top of the steps and reached the shrine, we talked of its history and meaning. Kibune is one of the most famous of the several shrines associated with water around Lake Biwa. Although the Shinto religion considers objects like water to be gods, the god of this shrine, Takaokami-no-kami, is not the god of water itself but the provider of water, the "god of the river headwaters," who offers rains in times of drought or reduces flooding in times of too much rain. From the beginning of the Heian Period (794 AD), when the capital of Japan was moved to Kyoto, the government has sent senior officials to this shrine to pray for rain, or for less rain, in times of human need. As at all Shinto shrines, there was a beautiful large basin of water with a bamboo ladle atop, and we rinsed our hands to purify ourselves before clapping loudly and praying to the god of the river headwaters.

Like streams in Japan and the United States where I have taught students and worked with colleagues, Kibune Stream that flows near the base of the long steps to the shrine has great meaning for Kawanabe and many Japanese ecologists. In the early 1940s a young aquatic entomologist named Tokichi Kani studied the assemblage of aquatic insects in this steep and swiftly flowing stream and influenced other early Japanese ecologists who worked with him. One of them created a theory of how species in ecological communities divide habitat and resources, a topic of intense interest to American and English ecologists working during the same period. In a tragic turn of world events, Kani was conscripted into the military and killed by American armed forces on a South Pacific island during World War II. Kawanabe himself had first visited this shrine seventy-two years earlier, at the age of eight, in 1940. Because of the influential research by Kani and others, Kibune Stream was the site of many studies and exercises in the field ecology course taught at Kyoto University after the war, and Kawanabe has fond memories of those excursions in the early 1950s. Standing in the warm soft evening breeze among the cedars, Yoshi and I felt the nurturing and inspiring influence of Japan's ecological history surrounding us as Kawanabe described his experiences and those of others that connected him to this stream and to this place.

But our meeting here was also a reunion of colleagues and friends across the span of the nearly twenty-five years during which I had visited and worked in Japan. Nakano's wife, Hiromi, and their eldest daughter Nana, had joined Yoshi and Jeremy and me, and Katsuki Nakai, for dinner with Hiroya Kawanabe at a famous restaurant on Kibune Stream. We descended the steps and entered the restaurant, leaving our shoes by the steps at the door in the traditional custom. We were ushered by waitresses in dark *yukata* robes to a large platform built above and over the stream, with *tatami* mats and traditional low Japanese tables. Our dinner was graced by the sounds and sights and cool mist created by water rushing over the riffles just upstream as we dined on many freshwater fishes, crustaceans, and other animals and plants that Japan's freshwaters offer.

Suddenly, at the end of our long and busy trip to Japan, I felt supremely happy, surrounded by our past, the present, and the future. Hiroya Kawanabe and I recalled many of the trials for Nakano and me and our colleagues, of the fieldwork and conducting the analyses and publishing the results afterward. We talked of history and literature and science, and of course the wonderful Japanese food and *sake*. Nana, having just finished her first year at college, looked bright and happy in starting her new-found life, and Yoshi and Katsuki Nakai talked to her like the supportive uncles they had become. Jeremy drank in the atmosphere of these scientists that he had grown to know well through his own trips to Japan. And Hiromi was the happiest that I had seen her in the twelve years since Shigeru was lost. I know that raising their three children had been daunting, and yet now so rewarding as she basked in their daughter's bright countenance. For all of us, life is that thing of which you make the best that you can. Like Debbie and Emily, and Ben and me in my family,

Hiromi and each around the table had succeeded in grasping and celebrating the best that life could offer them.

At the end of the evening, we hurried from the restaurant, saying our *Gochiso sama deshita* salutations (Thank you for the meal) and bowing deeply to our hosts. We needed to catch the quaint old-style Japanese train down the cedar-swathed river valley back to Kyoto, so that Yoshi could catch the last *shinkansen* (bullet) train back to his home in Nagoya. Sitting across from my Japanese and American friends and colleagues on the train, I reflected on how our lives flow in this river of time, and are joined in confluence with others along the way.

I knew this moment could not last, and that this might be the last time I would be able to travel again to Japan. But I also knew that we, and I, had done what we had hoped to achieve. I had traveled across an ocean many times to learn and teach about streams and science and life, with a man who had dedicated his life to creating a new vision for the way riverscapes and landscapes support each other. When his journey on this earth ended, we had returned to Japan to follow the clues he left us and discovered new truths that confirmed their worth. We had supported his family as best we could and mentored some of his former students and colleagues to help them achieve their own frontiers in science.

Nakano's passion was so great, his insights into nature so clear, and his striving for excellence so inspiring, that his influence on science and our lives has transcended his passing away. His life and work are joined in each of our efforts to understand the ecology and wonder that rests beyond the reflective boundary between forest and stream, and he walks in these places yet.

Notes

Notes to Introduction

My first research in graduate school, in the East Branch of the Au Sable River, Michigan, is described in: Fausch KD, White RJ. 1981. Competition between brook trout (*Salvelinus fontinalis*) and brown trout (*Salmo trutta*) for positions in a Michigan stream. *Canadian Journal of Fisheries and Aquatic Sciences* 38:1220–1227.

The endangerment of freshwater species is chronicled in: Wilcove DS, Master LL. 2005. How many endangered species are there in the United States? *Frontiers in Ecology and the Environment* 3:414–420; Dudgeon D, Arthington AH, Gessner MO, Kawabata Z, Knowler DJ, Lévêque C, Naiman RJ, Prieur-Richard A, Soto D, Stiassny MLJ, Sullivan CA. 2006. Freshwater biodiversity: Importance, threats, status and conservation challenges. *Biological Reviews* 81:163–182.

The quote from Aldo Leopold is from: Leopold A. 1953. The Round River. In Leopold LB, editor. *Round River: From the Journals of Aldo Leopold,* 165. Oxford University Press: New York, NY.

The metaphors in the final sentence stem from two poems, "On Marriage" and "On Children," in Gibran K. 1923. *The Prophet*, 15–18. Alfred A. Knopf: New York, NY.

Notes to Chapter 1: An Awakening

Trout and charr are members of the family that scientists call Salmonidae, which includes the salmon, trout, charr (or char), whitefish, and grayling. Charrs are members of the genus *Salvelinus*, but some species of charr are often called trout, so in this book I explain the difference for each species where this occurs. The classification and distribution of native charr in Hokkaido are described in: Kawanabe H, Yamazaki F, Noakes DLG, editors. 1989. Biology of charrs and masu salmon: Proceedings of the International Symposium on Charrs and Masu Salmon. *Physiology and Ecology Japan*, special volume 1; Fausch KD, Nakano S, Ishigaki K. 1994. Distribution of two congeneric charrs in streams of Hokkaido Island, Japan: Considering multiple factors across scales. *Oecologia* 100:1–12.

The history of the name of Dolly Varden charr is described in: Behnke RJ. 2002. *Trout and Salmon of North America.* The Free Press, Simon and Schuster, Inc.: New York, NY.

My first research in graduate school on replacement of native brook trout by nonnative brown trout in Michigan streams is described in: Fausch KD, White RJ. 1981. Competition between brook trout (*Salvelinus fontinalis*) and brown trout (*Salmo trutta*) for positions in a Michigan stream. *Canadian Journal of Fisheries and Aquatic Sciences* 38:1220–1227.

Shigeru Nakano's previous research on these charr is described in: Nakano S, Furukawa-Tanaka T. 1994. Intra- and interspecific dominance hierarchies and variation in foraging tactics of two species of stream-dwelling charrs. *Ecological Research* 9:9–20.

The field experiment and field comparative studies we conducted on Dolly Varden and whitespotted charr in Poroshiri Stream are described in: Fausch KD, Nakano S, Kitano S. 1997. Experimentally induced foraging mode shift by sympatric charrs in a Japanese mountain stream. *Behavioral Ecology* 8:414–420; Nakano S, Fausch KD, Kitano S. 1999. Flexible niche partitioning via a foraging mode shift: A proposed mechanism for coexistence in stream-dwelling charrs. *Journal of Animal Ecology* 68:1079–1092.

The symposium "The Biology of Charrs and Masu Salmon" held in Sapporo, Japan, October 1988, is described in: Noakes DLG. 1989. Symposium to be remembered. *Environmental Biology of Fishes* 24:313–317.

Dr. Hiroya Kawanabe's early influential paper in English is: Kawanabe H. 1969. The significance of social structure in production of the 'ayu,' *Plecoglossus altivelis*. In Northcote TG, editor. *The Symposium on Salmon and Trout in Streams*, 243–251. H. R. MacMillan Lectures in Fisheries. University of British Columbia: Vancouver, BC.

The paper I presented at the 1988 Charr Symposium is: Fausch KD. 1989. Do gradient and temperature affect distributions of, and interactions between, brook charr (*Salvelinus fontinalis*) and other resident salmonids in streams? *Physiology and Ecology Japan*, special volume 1:303–322.

Dr. Ishigaki's book that Nakano and he gave me is: Ishigaki K. 1984. *Iwana no nazo wo ou* (Exploring the Mystery of Charrs). Iwanami-shoten: Tokyo, Japan (in Japanese).

The history of the roots of biogeography and their influence on ecology is presented in: McIntosh RP. 1985. *The Background of Ecology: Concept and Theory.* Cambridge University Press: New York, NY.

Early influential studies on factors that affect the distribution of species in ecological communities by Connell and Hairston are: Connell JH. 1961. The influence of interspecific competition and other factors on the distribution of the barnacle *Chthamalus stellatus*. *Ecology* 42:710–723; Hairston NG. 1980. The experimental test of an analysis of field distributions: Competition in terrestrial salamanders. *Ecology* 61:817–826.

My early theory that Nakano used in the research he presented at the Charr Symposium is in: Fausch KD. 1984. Profitable stream positions for salmonids: Relating specific growth rate to net energy gain. *Canadian Journal of Zoology* 62:441–451.

The abstract for the paper that Nakano presented at the Charr Symposium is: Nakano S. 1989. Interspecific social interaction of Japanese charr (*Salvelinus leucomaenis*) and masu salmon (*Oncorhynchus masou masou*) in a mountain stream, Japan. *Physiology and Ecology Japan*, special volume 1:357. (Abstract only)

Papers that used and extended the theory that I developed on profitable feeding positions for trout and salmon include: Hughes FN, Dill LM. 1990. Position choice by drift feeding salmonids: Model and test for Arctic grayling (*Thymallus arcticus*) in subarctic mountain streams, interior Alaska. *Canadian Journal of Fisheries and Aquatic Sciences* 47:2039–2048; Urabe H, Nakajima M, Torao M, Aoyama T. 2010. Evaluation of habitat quality for stream salmonids based on a bioenergetics model. *Transactions of the American Fisheries Society* 139:1665–1676. My paper reviewing the history of this theory is: Fausch, KD. 2014. A historical perspective on drift foraging models for stream salmonids. *Environmental Biology of Fishes* 97:453–464.

The detailed underwater measurements of salmon and charr positions and contests that Nakano made by snorkeling are described in three papers: Nakano S. 1994. Variation in agonistic encounters in a dominance hierarchy of freely interacting red-spotted masu salmon (*Oncorhynchus masou ishikawae*). *Ecology of Freshwater Fish* 3:153–158; Nakano S. 1995. Competitive interactions for foraging microhabitats in a size-structured interspecific dominance hierarchy of two sympatric stream salmonids in a natural habitat. *Canadian Journal of Zoology* 73:1845–1854; Nakano S. 1995. Individual differences in resource use, growth and emigration under the influence of a dominance hierarchy in fluvial red-spotted masu salmon in a natural habitat. *Journal of Animal Ecology* 64:75–84.

The laboratory study of fish positions I conducted for my PhD research is: Fausch KD, White RJ. 1986. Competition among juveniles of coho salmon, brook trout, and brown trout in a laboratory stream, and implications for Great Lakes tributaries. *Transactions of the American Fisheries Society* 115:363–381.

The studies by Japanese scientists on the effects of releasing hatchery chum salmon in the North Pacific Ocean are in: Kaeriyama M. 1989. Aspects of salmon ranching in Japan. *Physiology and Ecology Japan*, special volume 1:625–638; Kaeriyama M, Seo H, Kudo H, Nagata M. 2012. Perspectives on wild and hatchery salmon interactions at sea, potential climate effects on Japanese chum salmon, and the need for sustainable salmon fishery management reform in Japan. *Environmental Biology of Fishes* 94:165–177.

Notes to Chapter 2: Exploring the Mystery of Charrs

The previous study of charr in Poroshiri Stream by Nakano and Tanaka is in: Nakano S, Furukawa-Tanaka T. 1994. Intra- and interspecific dominance hierarchies and variation in foraging tactics of two species of stream-dwelling charrs. *Ecological Research* 9:9–20.

The evolution and dispersal of brown bears is described in: Miller CR, Waits LP, Joyce P. 2006. Phylogeography and mitochondrial diversity of extirpated brown bear (*Ursus arctos*) populations in the contiguous United States and Mexico. *Molecular Ecology* 15:4477–4485.

Working alone in the field is never recommended, for safety reasons, and soon after we started in 1991 Nakano was able to enlist more volunteers so that neither of us had to work alone.

Ishigaki's book describing charr distributions in Hokkaido is: Ishigaki K. 1984. *Iwana no nazo wo ou* (Exploring the Mystery of Charrs). Iwanami-shoten: Tokyo, Japan (in Japanese).

In addition to the factors I list that can drive the distribution of species, species can also be limited by predators, diseases, or parasites, or even the lack of other species that provide benefits (called "mutualists"). However, the focus of our research on the distributions of these two similar charr species was on competitive interactions.

The ideas underpinning condition-specific competition are described in a paper by Taniguchi and Nakano that reports the results of their laboratory experiments on competition between whitespotted and Dolly Varden charr: Taniguchi Y, Nakano S. 2000. Condition-specific competition: Implications for the altitudinal distribution of stream fishes. *Ecology* 81:2027–2039.

In addition to Ishigaki's (1984) book, detailed information and analysis of charr distributions in Hokkaido (see figure 4) is presented in: Fausch KD, Nakano S, Ishigaki K. 1994. Distribution of two congeneric charrs in streams of Hokkaido Island, Japan: Considering multiple factors across scales. *Oecologia* 100:1–12.

The evolution and distribution of charr in Japan is described in several papers and was clarified by personal communication with Dr. Shoichiro Yamamoto, National Research Institute of Agriculture, Fisheries Research Agency, Nikko, Japan. See: Yamamoto S, Morita K, Kitano S, Watanabe K, Koizumi I, Maekawa K, Takamura K. 2004. Phylogeography of white-spotted charr (*Salvelinus leucomaenis*) inferred from mitochondrial DNA sequences. *Zoological Science* 21:229–240;Yamamoto S, Maekawa K, Morita K, Crane P, Oleinik A. [in press] Phylogeography of a salmonid fish, Dolly Varden *Salvelinus malma*: Multiple glacial refugia in the North Pacific rim. *Zoological Science.*

I reviewed the types of experiments and evidence needed to test for interspecific competition between different species of fish in two papers: Fausch KD. 1988. Tests of competition between native and introduced salmonids in streams: What have we learned? *Canadian Journal of Fisheries and Aquatic Sciences* 45: 2238–2246; Fausch KD. 1998. Interspecific competition and juvenile Atlantic salmon: On testing effects and evaluating the evidence across scales. *Canadian Journal of Fisheries and Aquatic Sciences* 55 (suppl. 1): 218–231.

Although my plan to remove all fish from pools and restock them with specific numbers of the two charr species seems foolhardy to me now, at this time ecologists who studied stream fish thought that most fish stayed put and moved relatively little. We were just discovering in another research project that many did not, as described in chapter 3. See: Gowan C, Young MK, Fausch KD, Riley SC. 1994. Restricted movement in resident stream salmonids: A paradigm lost? *Canadian Journal of Fisheries and Aquatic Sciences* 51:2626–2637.

The field study that Nakano led, on the feeding behavior of charr in Poroshiri Stream, is described in: Nakano S, Fausch KD, Kitano S. 1999. Flexible niche partitioning via a foraging mode shift: A proposed mechanism for coexistence in stream-dwelling charrs. *Journal of Animal Ecology* 68:1079–1092.

The statement that persistence and determination alone are omnipotent came from a quote in a program for a memorial service for Calvin Coolidge (1933). See: *The Oxford Dictionary of Quotations.* 1999. Oxford University Press: Oxford, UK.

Our discoveries about factors that explain the charr distributions in Hokkaido are presented in: Fausch KD, Nakano S, Ishigaki K. 1994. Distribution of two congeneric charrs in streams of Hokkaido Island, Japan: Considering multiple factors across scales. *Oecologia* 100:1–12.

The results from the field experiment where drifting insects were depleted are in: Fausch KD, Nakano S, Kitano S. 1997. Experimentally induced foraging mode shift by sympatric charrs in a Japanese mountain stream. Behavioral *Ecology* 8:414–420. The equal competitive ability of the two species at equal size in interspecific dominance hierarchies is briefly described in this paper, including reference to additional data that are unpublished.

The results of the laboratory experiments testing competition between the two charr at two different temperatures are described in: Taniguchi Y, Nakano S. 2000. Condition-specific competition: Implications for the altitudinal distribution of stream fishes. *Ecology* 81:2027–2039.

Notes to Chapter 3: Riverscapes: How Streams Work

The idea of riverscapes, and the analogy of attempting to appreciate a landscape painting by viewing only a few glimpses of it, are from: Fausch KD, Torgersen CE, Baxter CV, Li HW. 2002. Landscapes to riverscapes: Bridging the gap between research and conservation of stream fishes. *BioScience* 52:483–498.

Professor Noel Hynes championed the idea that streams are a product of their catchment (the English term for watershed), in: Hynes, HBN. 1975. The stream and its valley (Edgardo Baldi Memorial Lecture). *Verhandlungen—Internationale Vereinigung für Theoretische und Angewandte Limnologie* 19:1–15.

For an understanding of groundwater, and the entire hydrologic cycle, see: Dunne T, Leopold LB. 1978. *Water in Environmental Planning.* W.H. Freeman and Company: New York, NY.

The research by Michigan geologists on predicting trout populations from geology is reported in two publications: Hendrickson GE, Doonan CJ. 1972. *Hydrology and recreation on the cold-water rivers of Michigan's Southern Peninsula.* U.S. Geological Survey, Water Information Series 3: Lansing, MI; Hendrickson GE, Knutilla RL, Doonan CJ. 1973. *Hydrology and recreation on the cold-water rivers of Michigan's Upper Peninsula.* U.S. Geological Survey, Water Information Series 4: Lansing, MI.

The role of wildfires in the ecology of streams and fish is described in: Dunham JB, Rosenberger AE, Luce CH, Rieman BE. 2007. Influences of wildfire and channel reorganization on spatial and temporal variation in stream temperature and the distribution of fish and amphibians. *Ecosystems* 10:335–346.

A conceptual model of the cycle of natural disturbances like landslides and their key role in creating habitat in streams is presented in: Reeves GH, Benda LE, Burnett KM, Bisson PA, Sedell JR. 1995. A disturbance-based ecosystem approach to maintaining and restoring freshwater habitats of evolutionarily significant units of anadromous salmonids in the Pacific Northwest. *American Fisheries Society Symposium* 17:334–349.

Tansley's original paper coining the term ecosystem is: Tansley AG. 1935. The use and abuse of vegetational concepts and terms. *Ecology* 16:284–307

The annual budget of organic matter entering a small New Hampshire stream is presented in: Findlay SEG, Likens GE, Hedin I, Fisher SG, McDowell WH. 1997. Organic matter dynamics in Bear Brook, Hubbard Brook Experimental Forest, New Hampshire, USA. *Journal of the North American Benthological Society* 16:43–46.

The River Continuum Concept was originally presented in: Vannote RL, Minshall GW, Cummins KW, Sedell JR, Cushing CE. 1980. The river continuum concept. *Canadian Journal of Fisheries and Aquatic Sciences* 37: 130–137.

Challenges to the River Continuum Concept, and the value of the theory in addressing these exceptions, are described in two papers and a book: Winterbourn MJ, Rounick JS, Cowie B. 1981. Are New Zealand stream ecosystems really different? *New Zealand Journal of Marine and Freshwater Research* 15:321–328; Barmuta LA, Lake PS. 1982. On the value of the River Continuum Concept. *New Zealand Journal of Marine and Freshwater Research* 16:227–229; Allan JD, Castillo MM. 2007. *Stream Ecology: Structure and Function of Running Waters*, 2nd edition. Springer, Dordrecht: The Netherlands.

The use of stable isotopes to determine the sources of carbon in stream invertebrates is described in: Finlay, JC. 2001. Stable carbon isotope ratios of river biota: Implications for energy flow in lotic food webs. *Ecology* 84:1052–1064.

The role of bacteria in incorporating dissolved organic matter into the biofilm is reported in: Hall RO Jr., Meyer JL. 1998. The trophic significance of bacteria in a detritus-based stream food web. *Ecology* 79:1995–2012.

The preference of invertebrates for nutrient-rich food sources is described in: Cross WF, Benstead JP, Rosemond AD, Wallace JB. 2003. Consumer-resource stoichiometry in detritus-based streams. *Ecology Letters* 6:721–732.

Functional groups of insects, categorized by the roles they play, are described in: Vannote RL, Minshall GW, Cummins KW, Sedell JR, Cushing CE. 1980. The river continuum concept. *Canadian Journal of Fisheries and Aquatic Sciences* 37:130–137.

The reasons scientists have proposed for why aquatic invertebrates drift in streams are summarized in: Allan JD, Castillo MM. 2007. *Stream Ecology:*

Structure and Function of Running Waters, 2nd edition. Springer, Dordrecht: The Netherlands.

The role of terrestrial insects that fall into streams in feeding fish is reviewed in: Baxter CV, Fausch KD, Saunders WC. 2005. Tangled webs: Reciprocal flows of invertebrate prey link streams and riparian zones. *Freshwater Biology* 50:201–220.

The relative weight of aquatic insects versus terrestrial insects that fall into streams is presented in: Nakano S, Fausch KD, Kitano S. 1999. Flexible niche partitioning via a foraging mode shift: A proposed mechanism for coexistence in stream-dwelling charrs. *Journal of Animal Ecology* 68:1079–1092. The contribution of terrestrial insects to the annual energy budgets of trout and charr is described in: Nakano S, Murakami M. 2001. Reciprocal subsidies: Dynamic interdependence between terrestrial and aquatic food webs. *Proceedings of the National Academy of Sciences USA* 98:166–170.

The results of field experiments by Nakano and his colleagues using greenhouses are summarized in: Fausch KD, Baxter CV, Murakami M. 2010. Multiple stressors in north temperate streams: Lessons from linked forest-stream ecosystems in northern Japan. *Freshwater Biology* 55 (suppl. 1): 120–134.

The adage about the isolation of experts in narrow disciplines is attributed to Nicholas Murray Butler, a well-known American philosopher, diplomat, and educator who served as president of Columbia University for more than forty years (1902–1945) and won the Nobel Peace Prize.

The paper on emerging insects from a Kansas prairie stream that sparked Nakano's ideas is: Gray LJ. 1993. Response of insectivorous birds to emerging aquatic insects in riparian habitats of a tallgrass prairie stream. *American Midland Naturalist* 129:288–300.

The study by Nakano and Murakami on movements of insects between the stream and forest, and their use by birds and fish, is described in: Nakano S, Murakami M. 2001. Reciprocal subsidies: Dynamic interdependence between terrestrial and aquatic food webs. *Proceedings of the National Academy of Sciences USA* 98:166–170. Nakano's first greenhouse study is reported in: Nakano S, Miyasaka H, Kuhara N. 1999. Terrestrial-aquatic linkages: Riparian arthropod inputs alter trophic cascades in a stream food web. *Ecology* 80:2435–2441.

Our field experiment that influenced Nakano to use greenhouses to cut off movement of insects is: Fausch KD, Nakano S, Kitano S. 1997. Experimentally induced foraging mode shift by sympatric charrs in a Japanese mountain stream. *Behavioral Ecology* 8:414–420.

The studies on river-watershed linkages by Mary Power and John Sabo and their colleagues are reported in three publications: Power ME, Rainey WE, Parker MS, Sabo JL, Smyth A, Khandwala S, Finlay JC, McNeely FC, Marsee K, Anderson C. 2004. River to watershed subsidies in an old-growth conifer forest. In Polis GA, Power ME, Huxel GR, editors. *Food Webs at the Landscape Level*, 217–240. University of Chicago Press: Chicago, IL; Sabo JL, Power ME. 2002. Numerical response of lizards to aquatic insects and short-term consequences for terrestrial prey. *Ecology* 83:3023–3036; Sabo JL, Power ME. 2002. River-watershed exchange: Effects of riverine subsidies on riparian lizards and their terrestrial prey. *Ecology* 83:1860–1869.

The proportion of energy in a small New Hampshire stream contributed by fish and stream invertebrates is reported in: Fisher SE, Likens GE. 1973. Energy flow in Bear Brook, New Hampshire: An integrative approach to stream ecosystem metabolism. *Ecological Monographs* 43:421–439.

A summary of decades of work on lake food webs and their trophic cascades can be found in: Carpenter SR, Cole JJ, Kitchell JF, Pace ML. 2010. Trophic cascades in lakes: Lessons and prospects. In Terborgh J, Estes JA. *Trophic Cascades: Predators, Prey, and the Changing Dynamics of Nature*, 55–70. Island Press: Washington, DC.

A synthesis of the research on trophic cascades in aquatic and terrestrial ecosystems is: Eisenberg C. 2010. *The Wolf's Tooth: Keystone Predators, Trophic Cascades, and Biodiversity*. Island Press: Washington, DC.

Mary Power's 1989 field experiment demonstrating a trophic cascade in rivers is reported in: Power ME. 1990. Effects of fish in river food webs. *Science* 250:411–415. Shigeru Nakano's first field experiment with greenhouses that also demonstrated a trophic cascade is reported in: Nakano S, Miyasaka H, Kuhara N. 1999. Terrestrial-aquatic linkages: Riparian arthropod inputs alter trophic cascades in a stream food web. *Ecology* 80:2435–2441.

The linkages throughout watersheds, from tiny headwater rivulets that transport invertebrates and leaves downstream, to salmon that import marine-derived nitrogen from the ocean, are described in two papers: Holtgrieve GW, Schindler DE, Jewett PK. 2009. Large predators and biogeochemical hotspots: Brown bear (*Ursus arctos*) predation on salmon alters nitrogen cycling in riparian soils. *Ecological Research* 24:1125–1135; Wipfli MS, Baxter CV. 2010. Linking ecosystems, food webs, and fish production: Subsidies in salmonid watersheds. *Fisheries* 35:373–387.

The problem of understanding riverscapes based on glimpses of stream channel from one bend to the next is the opening premise of the conceptual paper by: Fausch KD, Torgersen CE, Baxter CV, Li HW. 2002. Landscapes to riverscapes: Bridging the gap between research and conservation of stream fishes. *BioScience* 52:483–498.

Early studies marking fish to measure their home ranges, and a critique of the conclusions drawn, are summarized in two papers: Gowan C, Young MK, Fausch KD, Riley SC. 1994. Restricted movement in resident stream salmonids: A paradigm lost? *Canadian Journal of Fisheries and Aquatic Sciences* 51:2626–2637; Fausch KD, Young MK. 1995. Evolutionarily significant units and movement of resident stream fishes: A cautionary tale. *American Fisheries Society Symposium* 17:360–370.

The results of our large-scale field experiment to measure the effects of log structures on trout abundance, and a companion study on trout movement, are described in two papers: Gowan C, Fausch KD. 1996. Long-term demographic responses of trout populations to habitat manipulation in six Colorado streams. *Ecological Applications* 6:931–946; Gowan C, Fausch KD. 1996. Mobile brook trout in two high-elevation Colorado streams: Reevaluating the concept of restricted movement. *Canadian Journal of Fisheries and Aquatic Sciences* 53:1370–1381.

Our study to measure the effects of the log structures twenty-one years later is described in: White SL, Gowan C, Fausch KD, Harris JG, Saunders WC. 2011. Response of trout populations in five Colorado streams two decades after habitat manipulation. *Canadian Journal of Fisheries and Aquatic Sciences* 68:2057–2063.

The history of logging and log drives and their effects on streams in this region is presented in: Fausch KD, Young MK. 2004. Interactions between forests and fish in the Rocky Mountains of the USA. In Northcote TG, Hartman GF, editors. *Fishes and Forestry: Worldwide Watershed Interactions and Management*, 463–484. Blackwell Science: Oxford, UK.

The paper we published critiquing trout movement studies is: Gowan C, Young MK, Fausch KD, Riley SC. 1994. Restricted movement in resident stream salmonids: A paradigm lost? *Canadian Journal of Fisheries and Aquatic Sciences* 51:2626–2637. The paper on use of habitat by fish in riverscapes is: Fausch KD, Torgersen CE, Baxter CV, Li HW. 2002. Landscapes to riverscapes: Bridging the gap between research and conservation of stream fishes. *BioScience* 52:483–498; The use of the riverscape concept in stream fish ecology is described in: Fausch KD. 2010. A renaissance in stream fish ecology. *American Fisheries Society Symposium* 73:199–206.

The different theories and evidence that scientists have generated about how fish use complementary habitats in riverscapes are reviewed in: Falke JA, Fausch KD. 2010. From metapopulations to metacommunities: Linking theory with empirical observations of the spatial population dynamics of stream fishes. *American Fisheries Society Symposium* 73:207–233.

Notes to Chapter 4: Tragedy in the Sea of Cortez

The events surrounding the accident that claimed the lives of Shigeru Nakano, Gary Polis, Masahiko Higashi, Takuya Abe, and Michael Rose were reassembled from my personal notes on a phone call from Yoshinori Taniguchi on March 31, 2000 from Mexico, and seven newspaper articles saved from the days after the accident (ordered here chronologically): *News UCDavis*. March 29, 2000. Three confirmed dead in UC Davis boat accident [www.news.ucdavis.edu]; Hendrix A. March 29, 2000. At least 3 drown in UC Davis Baja trip. *SF Gate* [www.sfgate.com]; Fox B. March 29, 2000. At least 3 die in capsizing of research boat off Baja: UC Davis spider expert among those missing. *Santa Barbara News-Press* [Associated Press]; *UCDavis News and Information*. March 30, 2000. Research expedition survivors' return to Sacramento Airport news conference; March 30, 2000 [www.news.ucdavis.edu]; Fox B. March 30, 2000. Missing professors are feared dead at sea. *Santa Barbara News-Press* [Associated Press]; Hardy T, Wiegand S, Furillo A. March 31, 2000. UCD scientists were heroes, Baja survivors say. *Sacbee* [*Sacramento Bee*] *News* (www.Sacbee.com); Kerr J. March 31, 2000. Survivors return to university after harrowing accident. *SF Gate* [www.sfgate.com].

Three papers that describe the research conducted by Gary Polis and his colleagues in the Sea of Cortez are: Polis GA, Hurd SD. 1996. Allochthonous input across habitats, subsidized consumers, and apparent trophic cascades: Examples from the ocean-land interface. In Polis GA, Winemiller KO, editors. *Food Webs: Integration of Patterns and Dynamics*, 275–285. Kluwer Academic Publishers: Boston, MA; Polis GA, Hurd SD. 1996. Linking marine and terrestrial food webs: Allochthonous input from the ocean supports high secondary productivity on small islands and coastal land communities. *American Naturalist* 147:396–423; Polis GA, Sánchez-Piñero F, Stapp PT, Anderson WB, Rose MD. 2004. Trophic flows from water to land: Marine input affects food webs of island and coastal ecosystems worldwide. In Polis GA, Power ME, and Huxel GR, editors. *Food Webs at the Landscape Level*, 200–216. University of Chicago Press: Chicago, IL.

Nakano's breakthrough paper is: Nakano S, Miyasaka H, Kuhara N. 1999. Terrestrial-aquatic linkages: Riparian arthropod inputs alter trophic cascades in a stream food web. *Ecology* 80:2435–2441.

Although newspapers reported their destination as Isla Cabeza de Caballo (Horsehead Island), because this is apparently where the boat carrying Nakano and Polis and others was later found, their destination was actually

Isla la Ventana (personal communication, Yoshinori Taniguchi, October 31, 2012). The boat may have drifted after capsizing in the accident.

My tribute to the Japanese scientists is: Fausch K. (unpublished) April 10, 2000. A tribute to Takuya Abe, Masahiko Higashi, and Shigeru Nakano. Delivered at the memorial ceremony "A Celebration of the Lives of Gary Polis, Michael Rose, Masahiko Higashi, Takuya Abe, and Shigeru Nakano." University of California, Davis. It was reported in: Anonymous [Associated Press]. April 11, 2000. Scientists who drowned honored at memorial. *San Francisco Chronicle.*

Nakano and Murakami's groundbreaking research is described in: Nakano S, Murakami M. 2001. Reciprocal subsidies: Dynamic interdependence between terrestrial and aquatic food webs. *Proceedings of the National Academy of Sciences USA* 98:166–170.

For an image of our final moments at CER, see Fausch KD. 2000. Shigeru Nakano: An uncommon Japanese fish ecologist. *Environmental Biology of Fishes* 59:359–364.

Three of the main papers from Nakano's research that I used to develop the grant proposal included: Nakano S, Miyasaka H, Kuhara N. 1999. Terrestrial-aquatic linkages: Riparian arthropod inputs alter trophic cascades in a stream food web. *Ecology* 80:2435–2441; Fausch KD, Nakano S, Kitano S. 1997. Experimentally induced foraging mode shift by sympatric charrs in a Japanese mountain stream. *Behavioral Ecology* 8:414–420; Nakano S, Kawaguchi Y, Taniguchi Y, Miyasaka H, Shibata Y, Urabe H, Kuhara N. 1999. Selective foraging on terrestrial invertebrates by rainbow trout in a forested headwater stream in northern Japan. *Ecological Research* 14:351–360.

The research using greenhouses by Kawaguchi, Taniguchi, and Nakano is described in: Kawaguchi Y, Nakano S, Taniguchi Y. 2003. Terrestrial invertebrate inputs determine the local abundance of stream fishes in a forested stream. *Ecology* 84:701–708.

The research by Kishi and his colleagues is reported in: Kishi D, Murakami M, Nakano S, Maekawa K. 2005. Water temperature determines strength of top-down control in a stream food web. *Freshwater Biology* 50:1315–1322. Mary Power's tribute to Kishi's research is in: Fausch KD, Power ME, Murakami M. 2002. Linkages between stream and forest food webs: Shigeru Nakano's legacy for ecology in Japan. *Trends in Ecology and Evolution* 17:429–434.

Notes to Chapter 5: Riverwebs

The design and results of the field experiment are described in: Baxter CV, Fausch KD, Murakami M, Chapman PL. 2004. Fish invasion restructures stream and forest food webs by interrupting reciprocal prey subsidies. *Ecology* 85:2656–2663.

The most recent Plinian eruption of Mt. Tarumae that produced large amounts of pumice was in 1739 (http://en.wikipedia.org/wiki/Mount_Tarumae).

The three papers that describe and review the results of our large-scale field experiment are: Baxter CV, Fausch KD, Murakami M, Chapman PL. 2004. Fish invasion restructures stream and forest food webs by interrupting reciprocal prey subsidies. *Ecology* 85:2656–2663; Baxter CV, Fausch KD, Murakami M, and Chapman PL. 2007. Invading rainbow trout usurp a terrestrial prey subsidy from native charr and reduce their growth and abundance. *Oecologia* 153:461–470; Fausch KD, Baxter CV, Murakami M. 2010. Multiple stressors in north temperate streams: Lessons from linked forest-stream ecosystems in northern Japan. *Freshwater Biology* 55 (suppl. 1): 120–134.

The fourth treatment in the experiment, with both the greenhouse and rainbow trout, had a very similar effect as the other two with either factor alone. The rainbow trout again usurped most terrestrial invertebrates, and Dolly Varden switched to eating more benthic invertebrates, which then declined compared with the control reaches. Periphyton bloomed, and insect emergence also declined, just like in the other treatments. We did not measure spiders for this treatment, because we reasoned that the result would be the same as for the reaches with greenhouses only. Greenhouses cut off insect emergence by the same amount whether or not rainbow trout were introduced into stream reaches.

Richard Levins' paper is: Levins R. 1966. The strategy of model building in population biology. *American Scientist* 54:421–431.

In 2001, the Ministry of Construction in Japan was combined with others to become the Ministry of Land, Infrastructure, Transport, and Tourism.

The design and results of the 2003 field research are described and reviewed in two papers: Baxter CV, Fausch KD, Murakami M, and Chapman PL. 2007. Invading rainbow trout usurp a terrestrial prey subsidy from native charr and reduce their growth and abundance. *Oecologia* 153:461–470; Fausch KD, Baxter CV, Murakami M. 2010. Multiple stressors in north temperate streams: Lessons from linked forest-stream ecosystems in northern Japan. *Freshwater Biology* 55 (suppl. 1): 120–134.

Sir Isaac Newton's idea is based on a remark in a letter to his rival Robert Hooke, February 15, 1676. See http://en.wikipedia.org/wiki/Isaac_Newton

For William Kittredge's idea, see: Kittredge W. 2007. Leaving the Ranch. In *The Next Rodeo: New and Selected Essays*, 89–101. Graywolf Press: Saint Paul, MN.

The International Fisheries Science Prize is awarded every four years at the World Fisheries Congress, for outstanding contributions to global fisheries science and fish conservation. The inaugural prize was awarded to me in 2008, based on nominations from the twelve organizations that belong to the World Council of Fisheries Societies, a nonprofit, nongovernmental organization that includes the American Fisheries Society, the Japanese Society of Fisheries Science, and the Fisheries Society of the British Isles.

Notes to Chapter 6: Running Dry

Wallace Stegner's work referred to here is: Stegner W. 1980. Introduction: Some geography, some history. In *The Sound of Mountain Water*, 9–38. E. P. Dutton. Reprinted 1997 by Penguin Books: New York, NY.

Julie Scheurer's research and our findings about brassy minnow ecology and distribution are described in two papers: Scheurer JA, Fausch KD, Bestgen KR. 2003. Multi-scale processes regulate brassy minnow persistence in a Great Plains river. *Transactions of the American Fisheries Society* 132:840–855; Scheurer JA, Bestgen KR, and Fausch KD. 2003. Resolving taxonomy and historic distribution for conservation of rare Great Plains fishes: *Hybognathus* (Teleostei: Cyprinidae) in eastern Colorado basins. *Copeia* 2003:1–12.

For Heraclitus's quote, see the epigraph to chapter 9.

The journals from John Frémont's expeditions are in: Jackson D, Spence ML. 1970. *The Expeditions of John Charles Frémont,* vol. 1. University of Illinois Press: Urbana.

The history of settling and farming the plains of eastern Colorado, and the resulting Dust Bowl, can be found in: Fardal L. 2003. Effects of groundwater pumping for irrigation on stream properties of the Arikaree River on the Colorado plains. Master of Science thesis. Department of Civil and Environmental Engineering, Colorado State University, Fort Collins, Colorado; Egan T. 2006. *The Worst Hard Time: The Untold Story of Those Who Survived the Great American Dust Bowl.* Houghton Mifflin Company: New York, NY.

The geological history of the region, and the zoogeography of its fishes, are in: Cross FB, Mayden RL, Stewart JD. 1986. Fishes in the western Mississippi drainage. In Hocutt CH, Wiley EO, editors. *The Zoogeography of North American Freshwater Fishes*, 363–412. John Wiley and Sons: New York, NY.

The history of the first fish collections in eastern Colorado basins is in: Fausch KD, Bestgen KR. 1997. Ecology of fishes indigenous to the central and southwestern Great Plains. In Knopf FL, Samson FB, editors. *Ecology and Conservation of Great Plains Vertebrates*, 131–166. Ecological Studies 125. Springer-Verlag: New York, NY.

The list of fishes originally found in the Arikaree River basin, and those lost to date, are in: Falke JA, Fausch KD, Magelky R, Aldred A, Durnford DS, Riley LK, Oad R. 2011. The role of groundwater pumping and drought in shaping ecological futures for stream fishes in a dryland river basin of the western Great Plains, USA. *Ecohydrology* 4:682–697.

Our research on fish movement in riverscapes is described in chapter 3 and summarized in two papers: Gowan C, Young MK, Fausch KD, Riley SC. 1994. Restricted movement in resident stream salmonids: A paradigm lost? *Canadian Journal of Fisheries and Aquatic Sciences* 51:2626–2637; Fausch, KD, Torgersen CE, Baxter CV, Li HW. 2002. Landscapes to riverscapes: Bridging the gap between research and conservation of stream fishes. *BioScience* 52:483–498.

Ted Labbe's research is described in: Labbe TR, Fausch KD. 2000. Dynamics of intermittent stream habitat regulate persistence of a threatened fish at multiple scales. *Ecological Applications* 10:1774–1791.

Jeff Falke's dissertation research on the Arikaree River and its fishes are recorded in four publications: Falke JA, Fausch KD, Bestgen KR, Bailey LL. 2010. Spawning phenology and habitat use in a Great Plains, USA, stream fish assemblage: An occupancy estimation approach. *Canadian Journal of Fisheries and Aquatic Sciences* 67:1942–1956; Falke JA, Bestgen KR, Fausch KD. 2010. Streamflow reductions and habitat drying affect growth, survival, and recruitment of brassy minnow across a Great Plains riverscape. *Transactions of the American Fisheries Society* 139:1566–1583; Falke JA, Fausch KD, Magelky R, Aldred A, Durnford DS, Riley LK, Oad R. 2011. The role of groundwater pumping and drought in shaping ecological futures for stream fishes in a dryland river basin of the western Great Plains, USA. *Ecohydrology* 4:682–697; Falke JA, Bailey LL, Fausch KD, Bestgen KR. 2012. Colonization and extinction in dynamic habitats: An occupancy approach for a Great Plains stream fish assemblage. *Ecology* 93:858–867.

The paper describing use of infrared videography to measure stream temperature from a helicopter is: Torgersen CE, Price DM, Li HW, McIntosh BA. 1999. Multiscale thermal refugia and stream habitat associations of Chinook salmon in northeastern Oregon. *Ecological Applications* 9:301–319.

The idea of metapopulations for these fish is described in: Falke JA, Fausch KD. 2010. From metapopulations to metacommunities: Linking theory with empirical observations of the spatial population dynamics of stream fishes. *American Fisheries Society Symposium* 73:207–233.

Research by Professor Ram Oad's student Lisa Fardal is described in: Fardal L. 2003. Effects of groundwater pumping for irrigation on stream properties of the Arikaree River on the Colorado plains. Master of Science thesis. Department of Civil and Environmental Engineering, Colorado State University, Fort Collins, Colorado.

Information on the Ogallala Aquifer in general, pumping in the Republican River basin, and the Republican River Compact among states was garnered from several publications, a seminar, personal communication from Robert Longenbaugh (retired, Colorado Assistant State Engineer, and Colorado State University), and theses of five graduate students working with professors Ram Oad and Deanna Durnford: Banning RO. 2010. Analysis of the groundwater/surface water interactions in the Arikaree River Basin of eastern Colorado. Master of Science thesis, Department of Civil and Environmental Engineering, Colorado State University, Fort Collins, Colorado; Colorado State Engineer's Office. 2010. Republican River Compact Fact Sheet (Oct 2010). Unpublished. Denver, CO; Dennehy KF. 2000. High Plains regional ground-water study: U.S. Geological Survey Fact Sheet FS-091-00. http://co.water.usgs.gov/nawqa/hpgw/factsheets/DENNEHYFS1.html; Magelky RD. 2010. Groundwater modeling and water balance methods for predicting stream habitat. Master of Science thesis, Department of Civil and Environmental Engineering, Colorado State University, Fort Collins, Colorado; Riley LK. 2009. Finding the balance: A case study of irrigation, riparian evapotranspiration, and hydrology of the Arikaree River basin. Master of Science thesis, Department of Civil and Environmental Engineering, Colorado State University, Fort Collins, Colorado; Squires A. 2007. Groundwater response functions and water balances for parameter estimation and stream habitat modeling. Master of Science thesis, Department of Civil and Environmental Engineering, Colorado State University, Fort Collins, Colorado; Wachob ED. 2006. Pumping, riparian evapotranspiration, and the Arikaree River, Yuma County, Colorado. Master of Science thesis, Department of Civil and Environmental Engineering, Colorado State University, Fort Collins, Colorado; Wolfe, D. Colorado State Engineer. Case study: The Republican River dispute. Interdisciplinary Water Resources Seminar, Colorado State University. October 11, 2010.

The value of agricultural sales in the Arikaree River basin was estimated from data in: Thorvaldson J, Pritchett J. 2006. Economic impact analysis of irrigated acreage in four river basins in Colorado. Colorado Water Resources

Research Institute Completion Report #207. Fort Collins, CO. http://cwi.colostate.edu/publications/cr/207.pdf

Estimates of the drying of streams in western Kansas were provided by Dr. Keith Gido, Division of Biology, Kansas State University, Manhattan, KS.

The Republican River Compact is described in: Republican River Compact Administration. 2009. Resolution by the Republican River Compact Administration regarding approval of Colorado's augmentation plan and related accounting procedures submitted under subsection III.B.1.k of the Final Settlement Stipulation. Available at: http://www.republicanriver.com/Pipeline/ColoradosProposedResolution/tabid/180/Default.aspx (downloaded July 26, 2013).

Notes to Chapter 7: Natives of the West

Information on the distribution of cutthroat trout and other trout and charr in the West, their evolution, classification, and the history of their discovery is largely from two publications by the late Dr. Robert Behnke, a close colleague who passed away in September 2013: Behnke RJ. 1992. Native trout of western North America. American Fisheries Society Monograph 6, Bethesda, MD; Behnke RJ. 2002. *Trout and Salmon of North America*. The Free Press, Simon and Schuster, Inc.: New York, NY.

Market fishing for cutthroat trout is recorded in: Reed EB. 1996. *Moraine Park boyhood*. Self-published manuscript, Fort Collins, CO. Accessed from the Estes Park Library, Estes Park, CO.

The divergence of cutthroat and rainbow trout ancestors is thought to have occurred in the late Pliocene Epoch, as described in: Behnke RJ. 1992. Native trout of western North America. American Fisheries Society Monograph 6, Bethesda, MD.

Bob Behnke began his graduate studies on classification of western trout at the University of California-Berkeley in 1957, and the history of his early work is chronicled in: Behnke RJ. 1998. Going home again: Revisiting native trout watersheds of the West. *Trout* (Winter 1998). Republished in Behnke RJ. 2007. *About Trout: The Best of Robert J. Behnke from* Trout *Magazine*. The Lyons Press: Guilford, CT.

The distribution and evolution of Darwin's finches are described in: Weiner J. 1995. *The Beak of the Finch: A Story of Evolution in Our Time*. Vintage Books: New York, NY.

The evolution and taxonomy of the different subspecies and lineages of cutthroat trout in the southern Rocky Mountains of Colorado, Wyoming, Utah, and New Mexico are currently the subject of much research and debate. For simplicity, the original subspecies names and ranges designated by Bob Behnke are used here. See: Metcalf JL, Pritchard VL, Silvestri SM, Jenkins JB, Wood JS, Cowley DE, Evans RP, Shiozawa DK, Martin AP. 2007. Across the Great Divide: Genetic forensics reveals misidentification of endangered cutthroat trout populations. *Molecular Ecology* 16:4445–4454; Metcalf JL, Love Stowell S, Kennedy CM, Rogers KB, McDonald D, Epp J, Keepers K, Cooper A, Austin JJ, Martin AP. 2012. Historical stocking data and 19th century DNA reveal human-induced changes to native diversity and distribution of cutthroat trout. *Molecular Ecology* 21:5194–5207; Bestgen KR, Rogers KB, Granger R. 2013. Phenotype predicts genotype for lineages of native cutthroat trout in the southern Rocky Mountains. Final Report to the U.S. Fish and Wildlife Service, Colorado Field Office, Denver Federal Center (MS 65412): Denver, CO.

The huge spawning runs of cutthroat trout from Yellowstone Lake are described in: Gresswell RE, Liss WJ, Larson GL. 1994. Life-history organization of Yellowstone cutthroat trout (*Oncorhynchus clarki bouvieri*) in Yellowstone Lake. *Canadian Journal of Fisheries and Aquatic Sciences* 51(S1):298–309; Koel TM, Bigelow PE, Doepke PD, Ertel BD, Mahony DL. 2005. Nonnative lake trout result in Yellowstone cutthroat trout decline and impacts to bears and anglers. *Fisheries* 30(11):10–19; Gresswell RE. 2011. Biology, status, and management of the Yellowstone cutthroat trout. *North American Journal of Fisheries Management* 31:782–812.

The effects of overfishing on a westslope cutthroat trout population is described in: Bjornn TC, Johnson TH, Thurow RF. 1977. Angling versus natural mortality in northern Idaho cutthroat trout populations. In Barnhart RA and Roelofs TD, editors. *Catch-and-Release Fishing as a Management Tool*, 89–98. Humboldt State University: Arcata, CA.

Effects of early mining, logging, sawmills, cattle grazing, and water diversions on streams and fish in the Colorado mountains are described in: Metcalf JL, Love Stowell S, Kennedy CM, Rogers KB, McDonald D, Epp J, Keepers K, Cooper A, Austin JJ, Martin AP. 2012. Historical stocking data and 19th century DNA reveal human-induced changes to native diversity and distribution of cutthroat trout. *Molecular Ecology* 21:5194–5207; Fausch KD, Young MK. 2004. Interactions between forests and fish in the Rocky Mountains of the USA. In Northcote TG, Hartman GF, editors. *Fishes and Forestry: Worldwide Watershed Interactions and Management*, 463–484. Blackwell Science: Oxford, U.K.; Wohl EE. 2001. *Virtual Rivers: Lessons from the Mountain Rivers of the Colorado Front Range*. Yale University Press: New Haven, CT.

The history of the earliest artificial propagation of trout in North America is in: Behnke R. 2003. About trout: A fishy "Whodunit?" *Trout* (Spring 2003):

54–56. Records of the earliest trout stocking in Colorado are in the Supplemental Materials to Metcalf JL, Love Stowell S, Kennedy CM, Rogers KB, McDonald D, Epp J, Keepers K, Cooper A, Austin JJ, Martin AP. 2012. Historical stocking data and 19th century DNA reveal human-induced changes to native diversity and distribution of cutthroat trout. *Molecular Ecology* 21:5194–5207.

The current distribution and purity of the various subspecies of cutthroat trout require much field research and analysis to assess. Westslope cutthroat trout are estimated to be pure in 37% of the stream miles in their native range (Shepard et al. 2005; see below) and Yellowstone cutthroat trout in 28% of theirs (Gresswell 2011), but the other inland subspecies have not fared as well. Lahontan cutthroat trout are present in only 11% of their historic range (although some of these populations are not pure; Young and Harig 2001), and pure Colorado River cutthroat trout remain in only 6% of their range (Hirsch et al. 2006). For others like Paiute (Behnke 2002) and greenback cutthroat trout (Metcalf et al. 2012) only a handful of known pure populations remain, in far less than 1% of the native range. See: Behnke RJ. 2002. *Trout and Salmon of North America*. The Free Press, Simon and Schuster, Inc.: New York, NY; Gresswell RE. 2011. Biology, status, and management of the Yellowstone cutthroat trout. *North American Journal of Fisheries Management* 31:782–812; Hirsch CL, Albeke SE, Nesler TP. 2006. Range-wide status of Colorado River cutthroat trout (*Oncorhynchus clarkii pleuriticus*): 2005. Colorado River Cutthroat Trout Conservation Team. Unpublished report. Available from Colorado Parks and Wildlife: Denver, CO; Metcalf JL, Love Stowell S, Kennedy CM, Rogers KB, McDonald D, Epp J, Keepers K, Cooper A, Austin JJ, Martin AP. 2012. Historical stocking data and 19th century DNA reveal human-induced changes to native diversity and distribution of cutthroat trout. *Molecular Ecology* 21:5194–5207; Shepard BB, May BE, Urie W. 2005. Status and conservation of westslope cutthroat trout within the western United States. *North American Journal of Fisheries Management* 25:1426–1440; Young MK, Harig AL. 2001. A critique of the recovery of greenback cutthroat trout. *Conservation Biology* 15:1574–1584.

The current distribution of cutthroat trout in small fragments of stream habitat in Colorado is described in: Harig AL, Fausch KD, Young MK. 2000. Factors influencing success of greenback cutthroat trout translocations. *North American Journal of Fisheries Management* 20:994–1004; Harig AL, Fausch KD. 2002. Minimum habitat requirements for establishing translocated cutthroat trout populations. *Ecological Applications* 12:535–551; Roberts JJ, Fausch KD, Peterson DP, Hooten MB. 2013. Fragmentation and thermal risks from climate change interact to affect persistence of native trout in the Colorado River basin. *Global Change Biology* 19:1383–1398.

Reintroductions of native cutthroat trout in Colorado are described in: Harig AL, Fausch KD, Young MK. 2000. Factors influencing success of greenback cutthroat trout translocations. *North American Journal of Fisheries Management* 20:994–1004; Harig AL, Fausch KD. 2002. Minimum habitat requirements for establishing translocated cutthroat trout populations. *Ecological Applications* 12:535–551.

The invasion of nonnative lake trout into Yellowstone Lake is described in: Koel TM, Bigelow PE, Doepke PD, Ertel BD, Mahony DL. 2005. Nonnative lake trout result in Yellowstone cutthroat trout decline and impacts to bears and anglers. *Fisheries* 30(11):10–19; Munro AR, McMahon TE, Ruzycki JR. 2005. Natural chemical markers identify source and date of introduction of an exotic species: Lake trout (*Salvelinus namaycush*) in Yellowstone Lake. *Canadian Journal of Fisheries and Aquatic Sciences* 62:79–87; Gresswell RE. 2011. Biology, status, and management of the Yellowstone cutthroat trout. *North American Journal of Fisheries Management* 31:782–812.

The research on effects of brook trout on native cutthroat trout, in Willow Creek and three other streams, is described in: Peterson DP, Fausch KD. 2003. Dispersal of brook trout promotes invasion success and replacement of native cutthroat trout. *Canadian Journal of Fisheries and Aquatic Sciences* 60:1502–1516; Peterson DP, Fausch KD, White GC. 2004. Population ecology of an invasion: Effects of brook trout on native cutthroat trout. *Ecological Applications* 14:754–772.

For the history of the first stocking of brook trout in Colorado, see: Kennedy CM. 2010. Weird Bear Creek: A history of a unique cutthroat trout population. U.S. Fish and Wildlife Service, Colorado Fish and Wildlife Conservation Assistance Office: Estes Park, CO; Supplemental Material (online) for Metcalf JL, Love Stowell S, Kennedy CM, Rogers KB, McDonald D, Epp J, Keepers K, Cooper A, Austin JJ, Martin AP. 2012. Historical stocking data and 19th century DNA reveal human-induced changes to native diversity and distribution of cutthroat trout. *Molecular Ecology* 21:5194–5207.

Behnke's reports of the speed of brook trout invasion are in: Behnke RJ. 1992. Native trout of western North America. American Fisheries Society Monograph 6, Bethesda, MD.

Three papers describing research on effects of cold water temperature on cutthroat trout fry are: Harig AL, Fausch KD. 2002. Minimum habitat requirements for establishing translocated cutthroat trout populations. *Ecological Applications* 12:535–551; Coleman MA, Fausch KD. 2007. Cold summer temperature limits recruitment of age-0 cutthroat trout in high-elevation Colorado streams. *Transactions American Fisheries Society* 136:1231–1244; Coleman MA, Fausch KD. 2007. Cold summer temperature regimes cause a recruitment bottleneck in age-0 Colorado River cutthroat trout reared in laboratory streams. *Transactions American Fisheries Society* 136:639–654.

Two papers describing options for managing nonnative trout invasions, with or without barriers, are: Peterson DP, Rieman BE, Dunham JB, Fausch KD, Young MK. 2008. Analysis of trade-offs between threats of invasion by non-native brook trout (*Salvelinus fontinalis*) and intentional isolation for native westslope cutthroat trout (*Oncorhynchus clarkii lewisi*). *Canadian Journal of Fisheries and Aquatic Sciences* 65:557–573; Fausch KD, Rieman BE, Dunham JB, Young MK, Peterson DP. 2009. The invasion versus isolation dilemma: Tradeoffs in managing native salmonids with barriers to upstream movement. *Conservation Biology* 23:859–870.

Strategies for controlling brook trout by removal are analyzed in: Peterson DP, Fausch KD, Watmough J, Cunjak RA. 2008. When eradication is not an option: Modeling strategies for electrofishing suppression of nonnative brook trout to foster persistence of sympatric native cutthroat trout in small streams. *North American Journal of Fisheries Management* 28:1847–1867

Charles Darwin's treatise on evolution by natural selection is: Darwin C. 1859. *On the Origin of Species by Means of Natural Selection, or the Preservation of Favoured Races in the Struggle for Life*. John Murray, London, U.K.

My theory on the multiple reasons explaining paradoxical trout invasions is in: Fausch KD. 2008. A paradox of trout invasions in North America. *Biological Invasions* 10: 685–701.

Alfred Russel Wallace's theory, complementary to Darwin's, is in: Wallace AR. 1858. On the tendency of varieties to depart indefinitely from the original type. *Proceedings of the Linnean Society of London* 3:53–62.

Common carp introduction and invasion in England is recorded in two publications: Lever C. 1977. *The Naturalized Animals of the British Isles*, 439–446. Hutchinson and Company: London; Maitland PS, Campbell RN. 1992. *Freshwater Fishes of the British Isles*. Harper Collins Publishers, London.

The early history and ecology of global biological invasions is described in: Elton CS. 1958. *The Ecology of Invasions by Animals and Plants*. John Wiley and Sons: New York, NY.

Records of early trout introductions to various countries worldwide are in: Behnke RJ. 2002. *Trout and Salmon of North America*. The Free Press, Simon and Schuster, Inc.: New York, NY; Fausch KD, Taniguchi Y, Nakano S, Grossman GD, Townsend CR. 2001. Flood disturbance regimes influence rainbow trout invasion success among five Holarctic regions. *Ecological Applications* 11:1438–1455; Fausch KD. 2007. Introduction, establishment and effects of non-native salmonids: Considering the risk of rainbow trout invasion in the United Kingdom. *Journal of Fish Biology* 71(supplement D): 1–32; McDowall RM.1994. *Gamekeepers for the Nation: The Story of New Zealand's Acclimatization Societies 1861–1990.* Canterbury University Press: Christchurch, New Zealand; Worthington EB. 1940. Rainbows: A report on attempts to acclimatize rainbow trout in Britain. *Salmon and Trout Magazine* 100:241–260, and 101:62–99.

Invasion by nonnative rainbow trout into the native range of brook trout in the southern Appalachian Mountains is described in: Fausch KD. 2008. A paradox of trout invasions in North America. *Biological Invasions* 10: 685–701; Fausch KD. 1988. Tests of competition between native and introduced salmonids in streams: What have we learned? *Canadian Journal of Fisheries and Aquatic Sciences* 45:2238–2246.

Flow regime, emergence timing, and other factors that influence where rainbow and brook trout invade successfully worldwide are analyzed in: Fausch KD, Taniguchi Y, Nakano S, Grossman GD, Townsend CR. 2001. Flood disturbance regimes influence rainbow trout invasion success among five Holarctic regions. *Ecological Applications* 11:1438–1455; Fausch KD. 2008. A paradox of trout invasions in North America. *Biological Invasions* 10: 685–701. Emergence timing of brook trout in their native range, and brook and brown trout in Rocky Mountain streams is reported in: Fausch KD, White RJ. 1986. Competition among juveniles of coho salmon, brook trout, and brown trout in a laboratory stream, and implications for Great Lakes tributaries. *Transactions of the American Fisheries Society* 115:363–381; Latterell JJ, Fausch KD, Gowan C, Riley SC. 1998. Relationship of trout recruitment to snowmelt runoff flows and adult trout abundance in six Colorado mountain streams. *Rivers* 6:240–250.

The rise of modern humans and extinction of the Neanderthal species is described in: Green RE, and 55 coauthors. 2010. A draft Neanderthal genome. *Science* 328:710–722; Stringer C. 2013. *Lone Survivors: How We Came to Be the Only Humans on Earth.* St. Martin's Griffin: New York, NY.

Experiments on the outcome of competition and predation between brook and cutthroat trout fry are described in: Novinger DC. 2000. Reversals in competitive ability: Do cutthroat trout have a thermal refuge from competition with brook trout? Doctoral dissertation. Department of Zoology and Physiology, University of Wyoming: Laramie, WY.

The introduction of nonnative species into New Zealand, and their effects on native fishes in New Zealand and Australia are reviewed in: Crowl TA, Townsend CR, McIntosh AR. 1992. The impact of introduced brown and rainbow trout on native fish: The case of Australasia. *Reviews in Fish Biol-*

ogy and Fisheries 2:217–241; McDowall RM. 1994. *Gamekeepers for the Nation: The Story of New Zealand's Acclimatization Societies 1861–1990.* Canterbury University Press: Christchurch, New Zealand.

The introduction of nonnative trout into Yellowstone National Park is reported in: Rahel FJ. 1997. From Johnny Appleseed to Dr. Frankenstein: Changing values and the legacy of fisheries management. *Fisheries* 22(8): 8–9.

Trout fishing in Colorado by President Dwight Eisenhower is reported in: Norgren CA. 1999. Carl A. Norgren Papers: 1948–1964. Dwight D. Eisenhower Library, Abilene, Kansas, Accession A98-17. Accessed November 20, 2012.

Homogenization of fishes across the globe by stocking is described in: Rahel FJ. 2007. Biogeographic barriers, connectivity and homogenization of freshwater faunas: It's a small world after all. *Freshwater Biology* 52: 696–710.

The replacement of native cutthroat by rainbow trout in the Gunnison River, Colorado, is reported in: Behnke RJ. 2002. *Trout and Salmon of North America*, 192. The Free Press, Simon and Schuster, Inc.: New York, NY.

The argument that restoring native cutthroat trout to many streams is not worth the cost is in: Quist MC, Hubert WA. 2004. Bioinvasive species and the preservation of cutthroat trout in the western United States: Ecological, social, and economic issues. *Environmental Science and Policy* 7:303–313.

The logic and results from the field comparative study and field experiment supporting the claim that brook trout do have a different effect on stream-riparian ecosystems than native cutthroat trout is presented in three papers: Benjamin JR, Fausch KD, Baxter CV. 2011. Species replacement by a nonnative salmonid alters ecosystem function by reducing prey subsidies that support riparian spiders. *Oecologia* 167:503–512; Benjamin JR, Lepori F, Baxter CV, Fausch KD. 2013. Can replacement of native by non-native trout alter stream-riparian food webs? *Freshwater Biology* 58:1694–1709; Lepori F, Benjamin JR, Fausch KD, Baxter CV. 2012. Are invasive and native trout functionally equivalent predators? Results and lessons from a field experiment. *Aquatic Conservation: Marine and Freshwater Ecosystems* 22:787–798.

The use of insects emerging from streams by birds in the riparian zone is reported in two papers: Nakano S, Murakami M. 2001. Reciprocal subsidies: Dynamic interdependence between terrestrial and aquatic food webs. *Proceedings of the National Academy of Sciences USA* 98:166–170; Uesugi A, Murakami M. 2007. Do seasonally fluctuating aquatic subsidies influence the distribution pattern of birds between riparian and upland forests? *Ecological Research* 22:274–281.

Losses of cutthroat trout caused by invading lake trout in Yellowstone Lake that reduced prey for birds and mammals are reported in: Koel TM, Bigelow PE, Doepke PD, Ertel BD, Mahony DL. 2005. Nonnative lake trout result in Yellowstone cutthroat trout decline and impacts to bears and anglers. *Fisheries* 30 (11):10–19; Gresswell RE. 2011. Biology, status, and management of the Yellowstone cutthroat trout. *North American Journal of Fisheries Management* 31:782–812. Effects of such losses of salmonids in another ecosystem, and the importance of salmonid carcasses to supply nutrients that fertilize the riparian zone, are in: Spencer CN, McClelland BR, Stanford JA. 1991. Shrimp stocking, salmon collapse, and eagle displacement: Cascading interactions in the food web of a large aquatic ecosystem. *BioScience* 41:14–21; Helfield JM, Naiman RJ. 2001. Effects of salmon-derived nitrogen on riparian forest growth and implications for stream productivity. *Ecology* 82:2403–2409.

Effects of the loss of cutthroat trout from Yellowstone Lake on economic and social values are evaluated in: Gresswell RE, Liss WJ. 1995. Values associated with management of the Yellowstone cutthroat trout in Yellowstone National Park. *Conservation Biology* 9:159–165.

Research analyzing the potential for climate change to combine with other stressors and extirpate populations of native cutthroat trout is described in: Isaak DJ, Muhlfeld CC, Todd AS, Al-Chokhachy R, Roberts JJ, Kershner JL, Fausch KD, Hostetler SW. 2012. The past as prelude to the future for understanding 21st century climate effects on Rocky Mountain trout. *Fisheries* 37:542–556; Roberts JJ, Fausch KD, Peterson DP, Hooten MB. 2013. Fragmentation and thermal risks from climate change interact to affect persistence of native trout in the Colorado River basin. *Global Change Biology* 19:1383–1398; Wenger SJ, Isaak DJ, Luce CH, Neville HM, Fausch KD, Dunham JB, Dauwalter DC, Young MK, Elsner MM, Rieman BE, Hamlet AF, Williams JE. 2011. Flow regime, temperature and biotic interactions drive differential declines of trout species under climate change. *Proceedings of the National Academy of Sciences* 108:14175–14180.

The lower invasion success of brook trout in the northern Rocky Mountains is discussed in: Fausch KD. 2008. A paradox of trout invasions in North America. *Biological Invasions* 10: 685–701; Wenger SJ, Isaak DJ, Luce CH, Neville HM, Fausch KD, Dunham JB, Dauwalter DC, Young MK, Elsner MM, Rieman BE, Hamlet AF, Williams JE. 2011. Flow regime, temperature and biotic interactions drive differential declines of trout species under climate change. *Proceedings of the National Academy of Sciences* 108:14175–14180.

The success of angling regulations at increasing cutthroat trout in north Idaho streams is reported in: Bjornn TC, Johnson TH, Thurow RF. 1977. Angling versus natural mortality in northern Idaho cutthroat trout populations. In

Barnhart RA and Roelofs TD, editors. *Catch-and-Release Fishing as a Management Tool*, 89–98. Humboldt State University: Arcata, CA.

Bob Behnke's legacy of championing the understanding and conservation of native cutthroat trout is reviewed in two tributes: Fausch K, Bestgen K. 2014. The Behnke legacy. *Trout* (Winter 2014): 39; Piccolo JJ. 2014. Learning "About Trout": A student's tribute to Robert J. Behnke. *Fish and Fisheries* (in press).

Notes to Chapter 8: For the Love of Rivers

The epigraph is from: Leopold A. 1949. Song of the Gavilan. In *A Sand County Almanac*, 154. Oxford University Press: Oxford, UK.

Norman Maclean's novel is: Maclean N. 1976. *A River Runs Through It and Other Stories*. University of Chicago Press: Chicago, IL.

Wallace Stegner's tribute to Normal Maclean is: Stegner W. 1992. Haunted by waters: Norman Maclean. In *Where the Bluebird Sings to the Lemonade Springs: Living and Writing in the West*, 190–198. Random House, Inc.: New York, NY.

The percentage of U.S. streams and rivers in poor condition is from American Rivers. www.americanrivers.org. Accessed July 8, 2012.

The numbers of dams in the U.S. and Hokkaido can be found in: U.S. Army Corps of Engineers. 2006. National inventory of dams. http://crunch.tec.army.mil/nid/webpages/nid.cfm (accessed April 2006); Fukushima M, Kameyama S, Kaneko M, Nakao K, Steel EA. 2007. Modeling the effects of dams on freshwater fish distributions in Hokkaido, Japan. *Freshwater Biology* 52:1511–1524.

The biodiversity of freshwater fishes and other species is inventoried in: Dudgeon D, Arthington AH, Gessner MO, Kawabata Z, Knowler DJ, Lévêque C, Naiman RJ, Prieur-Richard A, Soto D, Stiassny MLJ, Sullivan CA. 2006. Freshwater biodiversity: Importance, threats, status and conservation challenges. *Biological Review* 81:163–182; Lévêque C, Oberdorff T, Paugy D, Stiassny MLJ, Tedesco PA. 2008. Global diversity of fish (Pisces) in freshwater. *Hydrobiologia* 595:545–567.

The imperilment of species, the rate of loss, and the tipping point for species extinction are summarized in: Wilson EO. 1993. Biophilia and the conservation ethic. In S. R. Kellert and E. O. Wilson, editors. *The Biophilia Hypothesis*, 31–41. Island Press, Washington, DC; Dirzo R, Raven PH. 2003. Global state of biodiversity and loss. *Annual Review of Environment and Resources* 28:137–167; Wilcove DS, Master LL. 2005. How many endangered species are there in the United States? *Frontiers in Ecology and the Environment* 3:414–420.

The use of fish and other freshwater organisms as indicators of degradation is reviewed in: Karr JR, Fausch KD, Angermeier PL, Yant PR, Schlosser IJ. 1986. Assessment of biological integrity in running waters: A method and its rationale. *Illinois Natural History Survey Special Publication* 5, Champaign, IL; Fausch KD, Lyons JD, Angermeier PL, Karr JR. 1990. Fish communities as indicators of environmental degradation. *American Fisheries Society Symposium* 8:123–144; Karr JR, Chu EW. 1998. *Restoring Life in Running Waters: Better Biological Monitoring*. Island Press: Washington, DC.

Aldo Leopold's quotes about damage to ecosystems being invisible is from: Leopold A. 1953. The Round River. In Leopold LB, editor. *Round River: From the Journals of Aldo Leopold*, 165. Oxford University Press: Oxford, UK.

Statistics on the sources of freshwater and their uses in the U.S. are from: Center for Sustainable Systems. 2011. U.S. water supply and distribution fact sheet. Publ. No. CSS05-17, University of Michigan: Ann Arbor, MI.

Data on the importance of fish for food protein are from: Helfman GS. 2007. *Fish Conservation: A Guide to Understanding and Restoring Global Aquatic Biodiversity and Fishery Resources*. Island Press: Washington, DC.

Information on the size of the Columbia River, its salmon runs, and fish catches in the Illinois River are in: Karr JR, Toth LA, Dudley DR. 1985. Fish communities of midwestern rivers: A history of degradation. *BioScience* 35:90–95; Northwest Power and Conservation Council (NPCC). 2009. Columbia River Basin Fish and Wildlife Program. NPCC Document 2009-09. Portland, OR; U.S. Geological Survey. 1990. Largest rivers in the United States. Open File Report 87–242. Washington, DC.

Research and writings on human preferences for natural and artificial sounds, including running water, are in: Carles JL, Barrio IL, Vincent de Lucio J. 1999. Sound influence on landscape values. *Landscape and Urban Planning* 43:191–200; Yang W, Kang J. 2005. Soundscape and sound preferences in urban squares: A case study in Sheffield. *Journal of Urban Design* 10:61–80; Jeon JY, Lee PJ, You J, Kang J. 2010. Perceptual assessment of quality of urban soundscapes with combined noise sources and water sounds. *Journal of the Acoustical Society of America* 127:1357–1366; Strand G. 2011. Ear-opening: A menagerie of musical performances tuned to the earth. *Orion* May-June 2011: 72–74.

The evolutionary basis for our innate preference for streams and other water features is described in: Wilson EO. 1984. *Biophilia: The Human Bond with*

Other Species. Harvard University Press: Cambridge, MA; Orians GH, Heerwagen JH. 1992. Evolved responses to landscapes. In Barkow JH, Cosmides L, Tooby J, editors. *The Adapted Mind: Evolutionary Psychology and the Generation of Culture*, 463–484. Oxford University Press: New York, NY; Ulrich RS. 1993. Biophilia, biophobia, and natural landscapes. In Kellert SR, Wilson EO, editors. *The Biophilia Hypothesis*, 73–137. Island Press: Washington, DC.

Evolution of modern humans in Africa and the spread to other continents is reviewed in: Stringer CB, Andrews P. 1988. Genetic and fossil evidence for the origin of modern humans. *Science* 239:1263–1268; Stringer C. 2012. *Lone Survivors: How We Came to Be the Only Humans on Earth*. St. Martin's Griffin: New York, NY.

The benefits to health, well-being, and creativity of viewing natural landscapes, and their water features, are reviewed in: Ulrich RS. 1993. Biophilia, biophobia, and natural landscapes. In Kellert SR, Wilson EO, editors. *The Biophilia Hypothesis*, 73–137. Island Press: Washington, DC.

John Daniel's quote on the difference between the character of lakes and rivers, and Stephen Kaplan's ideas about the preference for landscapes with mystery are from: Daniel J. 2009. Beginnings. In *The Far Corner: Northwestern Views on Land, Life, and Literature*, 49–56. Counterpoint Press: Berkeley, CA; Kaplan S. 1992. Environmental preference in a knowledge-seeking, knowledge-using organism. In Barkow JH, Cosmides L, and Tooby J, editors. *The Adapted Mind: Evolutionary Psychology and the Generation of Culture*, 581–598. Oxford University Press, New York.

The density of rivers across landscapes was calculated from bifurcation-ratio relationships in: Leopold LB, Wolman MG, Miller JP. 1964. *Fluvial Processes in Geomorphology*. W. H. Freeman and Co.: San Francisco, CA; Leopold LB. 1994. *A View of the River*. Harvard University Press: Cambridge, MA.

Richard Louv's ideas about the importance of children's play in the natural world are from: Louv R. 2005. *Last Child in the Woods: Saving our Children from Nature-Deficit Disorder*. Algonquin Books of Chapel Hill: Chapel Hill, NC.

The waterfalls and trees at the 9/11 Memorial in New York City are described on the 9/11 Memorial webpage—http://www.911memorial.org/memorial (accessed July 3, 2012). The idea that natural landscapes can help humans find deeper meaning and achieve better mental health is from: Ulrich RS. 1993. Biophilia, biophobia, and natural landscapes. In Kellert SR, Wilson EO, editors. *The Biophilia Hypothesis*, 73–137. Island Press: Washington, DC.

Philosophers label the tangible and intangible benefits that nature provides to humans instrumental values, and the benefits for which we receive no personal gain intrinsic values. See: Moore KD. 2007. In the shadow of the cedars: The spiritual values of old-growth forests. *Conservation Biology* 21:1120–1123.

Aldo Leopold's quote on conservation is from: Leopold A. 1953. Conservation. In Leopold LB, editor. *Round River: From the journals of Aldo Leopold*, 155. Oxford University Press: Oxford, UK.

The comment on grace and style from Gary Snyder is in: Dodge J. 1999. Foreword by Jim Dodge. In *The Gary Snyder Reader: Prose, Poetry, and Translations 1952–1998*, xv–xx. Counterpoint: Washington, DC. The comment about doing otherwise than striving is from Bruce Rieman (personal communication, August 1, 2012).

Aldo Leopold's quote about an ethical relation to land is from: Leopold A. 1949. The Land Ethic. In *A Sand County Almanac*, 223. Oxford University Press: Oxford, UK.

Kathleen Dean Moore's ideas on love of place are from: Moore KD. 2004. What it means to love a place. In *Pine Island Paradox*, 35–36. Milkweed Editions: Minneapolis, MN.

The idea about the Greek *agape* form of love is from Dr. David O'Hara, Department of Religion, Philosophy, and Classics, Augustana College, Sioux Falls, SD (personal communication, March 1, 2013).

The ideas on an ecological ethic of care, moral ground, and the importance of considering love in conservation are from: Gould, SJ. 1991. Unenchanted evening. *Natural History* 100(9):4–10; Moore KD. 2004. Toward an ecological ethic of care. In *Pine Island Paradox*, 62–65. Milkweed Editions: Minneapolis, MN; Moore KD, Nelson MP, editors. 2010. *Moral Ground: Ethical Action for a Planet in Peril*. Trinity University Press: San Antonio, TX; Orr DW. 2011. Love. In *Hope is an Imperative: The Essential David Orr*, 30–34. Island Press: Washington, DC.

Aldo Leopold's metaphor of the landscape and riverscape as a great orchestra that plays the song of songs is from: Leopold A. 1949. Song of the Gavilan. In *A Sand County Almanac*, 153. Oxford University Press: Oxford, UK.

Notes to Chapter 9: What Is the Future?

The epigraph is from Harris W. no date. Heraclitus: The complete fragments. Translation and commentary, and the Greek text. http://community.middlebury.edu/~harris/Philosophy/heraclitus.pdf. Accessed March 2013.

The research by Dan Bottom and Kim Jones on the salmon in the Salmon River and its estuary is described in: Bottom DL, Jones KK, Cornwell TJ, Gray A, Simenstad CA. 2005. Patterns of Chinook salmon migration and residency in the Salmon River estuary (Oregon). *Estuarine, Coastal, and Shelf Science* 64:79–93; Jones KK, Cornwell TJ, Bottom DL, Stein S, Wellard Kelly H, Campbell LA. 2011. Recovery of wild coho salmon in the Salmon River Basin, 2008–10. Monitoring Program Report Number OPSW-ODFW-2011-10, Oregon Department of Fish and Wildlife: Salem, OR; Jones KK, Cornwell TJ, Bottom DL, Campbell LA, Stein S. 2014. The contribution of estuary-resident life histories to the return of adult *Oncorhynchus kisutch*. *Journal of Fish Biology* 85:52–80.

Data on population status of salmon in rivers along the Oregon coast like the Salmon River are given in: Oregon Department of Fish and Wildlife. 2005. 2005 Oregon native fish status report, Volume II: *Assessment methods and population results*. Oregon State Department of Fish and Wildlife, Fish Division: Salem, OR. http://www.dfw.state.or.us/fish/ONFSR/docs/final/02-fall-chinook/fc-methods-coast.pdf (downloaded Jan 21, 2014). Coho salmon declined to low abundance during the 1980s, owing to overfishing and poor ocean conditions, but have rebounded some since owing to better management and ending hatchery programs. Adult Chinook numbers steadily increased during this same period, but many of those that return were released from hatcheries (personal communication, Kim Jones and Dan Bottom). Therefore, the numbers of wild adult salmon returning must have declined.

Effects of hatcheries on salmon survival, and the federal listing of coho salmon along the Oregon coast are in: Chilcote MW, Goodson KW, Falcy MR. 2011. Reduced recruitment performance in natural populations of anadromous salmonids associated with hatchery-reared fish. *Canadian Journal of Fisheries and Aquatic Sciences* 68:511–522; U.S. Federal Register (February 11, 2008) Vol. 73, No. 28. 50 CFR Parts 223 and 226. Endangered and threatened species: Final threatened listing determination, final protective regulations, and final designation of critical habitat for the Oregon coast Evolutionarily Significant Unit of coho salmon.

Research on juvenile salmon life histories using otoliths and tagging is described in: Bottom DL, Jones KK, Cornwell TJ, Gray A, Simenstad CA. 2005. Patterns of Chinook salmon migration and residency in the Salmon River estuary (Oregon). *Estuarine, Coastal, and Shelf Science* 64:79–93; Volk EC, Bottom DL, Jones KK, Simenstad CA. 2010. Reconstructing juvenile Chinook life history in the Salmon River estuary, Oregon, using otolith microchemistry and microstructure. *Transactions of the American Fisheries Society* 139:535–549; Jones KK, Cornwell TJ, Bottom DL, Campbell LA, Stein S. 2014. The contribution of estuary-resident life histories to the return of adult *Oncorhynchus kisutch*. *Journal of Fish Biology* 85:52–80.

The importance of diverse life-history types to resilience of Pacific salmon is described in: Schindler DE, Hilborn R, Chasco B, Boatright CP, Quinn TP, Rogers LA, Webster MS. 2010. Population diversity and the portfolio effect in an exploited species. *Nature* 465:609–613; Bottom DL, Jones KK, Simenstad CA, Smith CL. 2011. Reconnecting social and ecological resilience in salmon ecosystems. In Bottom DL, Jones KK, Simenstad CA, Smith CL, Cooper R, editors. *Pathways to Resilience: Sustaining Salmon Ecosystems in a Changing World*, 3–38. Oregon Sea Grant. Oregon State University, Corvallis.

The rich aquatic biodiversity in the southern Appalachian Mountains, and the effects of mountaintop mining and stream creation are described in: Bernhardt ES, Palmer MA. 2011. The environmental costs of mountaintop mining valley fill operations for aquatic ecosystems of the central Appalachians. *Annals of the New York Academy of Sciences* 1223:39–57; Palmer MA, Bernhardt ES, Schlesinger WH, Eshleman KN, Foufoula-Georgiou E, Hendryx MS, Lemly AD, Likens GE, Loucks OL, Power ME, White PS, Wilcock PR. 2010. Mountaintop mining consequences. *Science* 327:148–149.

Although the scheme I pose to divert water from the South Platte River to the Arikaree River is technically feasible, all of the water in the South Platte Basin is spoken for by those who hold the water rights, so changing the use of this water would be difficult and expensive. For background, see: Freeman DM. 2010. *Implementing the Endangered Species Act on the Platte Basin Water Commons*. University of Colorado Press: Boulder, CO.

The poem quoted is by Robert Burns (1785) "To a Mouse, on Turning Her Up in Her Nest with the Plough." In *Poems, Chiefly in the Scottish Dialect*. Printed by John Wilson: Kilmarnock, Scotland.

John Frémont's journals are reprinted in: Jackson D, Spence ML. 1970. *The Expeditions of John Charles Frémont*. Vol. 1. University of Illinois Press: Urbana.

Human alteration of flows in the South Platte River, and other plains rivers in eastern Colorado, is described in: Fausch KD, Bestgen KR. 1997. Ecology of fishes indigenous to the central and southwestern Great Plains. In Knopf FL, Samson FB, editors. *Ecology and Conservation of Great Plains Vertebrates*, 131–166. Ecological Studies 125. Springer-Verlag: New York; Strange EM, Fausch KD, Covich AP. 1999. Sustaining ecosystem services in human-dominated watersheds: Biohydrology and ecosystem processes in the South Platte River basin. *Environmental Management* 24:39–54.

Ongoing plans to divert water from west to east across the Continental Divide in Colorado are described in: Windy Gap Firming Project, Colorado Trout Unlimited. http://www.coloradotu.org/windy-gap/ (accessed Jan 6, 2013).

Invasion of zebra mussels and fish via canals and other structures, and costs of damage, are detailed in: Marchetti MP, Light T, Moyle PB, Viers JH. 2004. Fish invasions in California watersheds: Testing hypotheses using landscape patterns. *Ecological Applications* 14:1507–1525; Strayer D. 2009. Twenty years of zebra mussels: Lessons from the mollusk that made headlines. *Frontiers in Ecology and the Environment* 7:135–141; Mari L, Bertuzzo E, Casagrandi R, Gatto M, Levin SA, Rodriguez-Iturbe I, Rinaldo A. 2011. Hydrologic controls and anthropogenic drivers of the zebra mussel invasion of the Mississippi-Missouri river system. *Water Resources Research* 47: W03523.

Aldo Leopold's quote on the force behind conservation is from: Leopold A. ca. 1946. Quote from "The Meaning of Conservation" (unpublished manuscript). In Meine C, Knight RL, editors. 1999. *The Essential Aldo Leopold: Quotations and Commentaries*, 309. University of Wisconsin Press: Madison, WI.

Victor Frankl's philosophy is presented in: Frankl VE. 1984. *Man's Search for Meaning: An Introduction to Logotherapy*, 3rd edition. Simon and Schuster, Inc.: New York, NY.

Kathleen Dean Moore's ideas on our moral responsibility to land and rivers are from: Moore KD. 2004. The world depends on this. In *Pine Island Paradox*, 115–119. Milkweed Editions: Minneapolis, MN; Moore KD. 2004. Toward an ecological ethic of care. In *Pine Island Paradox*, 60–67. Milkweed Editions: Minneapolis, MN; Moore KD, Nelson MP. 2010. Toward a global consensus for ethical action. In Moore KD, Nelson MP, editors. *Moral Ground: Ethical Action for a Planet in Peril*, xv–xxiv. Trinity University Press: San Antonio, TX.

Resources used per year for the average family, and Wendell Berry's quote about soil, are in: Robecheck C. 2003. A land and people index. In *Coming to Land in a Troubled World*, 120–126. The Trust for Public Land: San Francisco, CA; Berry W. 2012 (first published 1969). Some thoughts on citizenship and conscience in honor of Don Pratt. In *The Long-legged House*, 89–106. Counterpoint: Berkeley, CA.

Water use for generating thermoelectric power is described in: Kenny JF, Barber NL, Hutson SS, Linsey KS, Lovelace JK, Maupin MA. 2009. Estimated use of water in the United States in 2005. U.S. Geological Survey Circular 1344: Reston, VA; McDonald RI, Olden JD, Opperman JJ, Miller WM, Fargione J, Revenga C, Higgins JV, Powell J. 2012. Energy, water and fish: Biodiversity impacts of energy-sector water demand in the United States depend on efficiency and policy measures. *PLOS One* 7(11): e50219; See also: A coal-fired thermoelectric power plant: Georgia Power's Plant Scherer (http://ga.water.usgs.gov/edu/wupt-coalplant-diagram.html; accessed Jan 24, 2013).

The statistic on water use for irrigation is from: Center for Sustainable Systems. 2011. U.S. water supply and distribution fact sheet. Publication CSS05-17, University of Michigan: Ann Arbor.

Data on per-capita water use in my region and other countries are from: Australian Bureau of Statistics. 4610.0-Water Account, Australia, 2010–11 (released 11-27-2012). http:/www.abs.gov.au/AUSSTATS/abs@.nsf/Latestproducts (downloaded December 26, 2012); Instituto Nacional de Estadística (Spain). 2009. Notas de prensa. Encuesta sobre el suministro y saneamiento del agua. Año 2007. http://www.ine.es/prensa/np561.pdf (downloaded December 26, 2012); Robecheck C. 2003. A land and people index. In *Coming to Land in a Troubled World*, 120–126. The Trust for Public Land: San Francisco, CA; U.S. Army Corps of Engineers. 2008. Draft Environmental Impact Statement: Northern Integrated Supply Project for the Northern Colorado Water Conservancy District. USCOE Omaha District: Omaha, NE.

Uses of corn to produce ethanol are discussed in: Lott MC. 2011. The U.S. now uses more corn for fuel than for feed. Scientific American: Plugged In Blog (http://blogs.scientificamerican.com/plugged-in/2011/10/07/the-u-s-now-uses-more-corn-for-fuel-than-for-feed/; accessed Jan 24, 2013); Fred Allendorf writes of cultivating "mindfulness," used in the practice of Zen Buddhism, as a way to connect our actions to consequences, such as the action of turning on light switches to the consequences for rivers of the dams that produce the electricity, in: Allendorf FW. 1997. The conservation biologist as Zen student. *Conservation Biology* 11:1045–1046.

The history and current laws governing use of water and instream flows in Colorado are summarized in: Grantham J. 2011. Synopsis of Colorado water law. Colorado Division of Water Resources: Denver, CO. http://water.state.co.us/DWRIPub/DWR%20General%20Documents/SynopsisOfCOWaterLaw.pdf (downloaded Jan 8, 2013); History of water rights. Colorado Division of Water Resources, Colorado Department of Natural Resources. http://water.state.co.us/SurfaceWater/SWRights/Pages/WRHistory.aspx (accessed Jan 24, 2013); Merriman D, Janicki AM. Colorado's instream flow program – how it works and why it's good for Colorado. cwcb.state.co.us (downloaded Jan 8, 2013); Prior Appropriation law. Colorado Division of Water Resources, Colorado Department of Natural Resources. http://water.state.co.us/SurfaceWater/SWRights/Pages/PriorApprop.aspx (accessed Jan 24, 2013).

Estimated water use per 5 X 5 foot bale of hay is based on data in two publications from Texas. Forage production per season is 11,000 pounds/acre (Banta 2012; see below), and requires 24 inches of irrigation (Masoner et al. 2003; 2 acre-feet of water per acre, or 651,703 gallons). At 1,100 pounds/bale (Banta 2012) this yields 10 bales/acre, so each bale requires about 65,000 gallons of water to produce; Banta J. 2012. Bale weight: How important is it? AgriLife Extension bulletin E-319. Texas A&M University: College Station. http://AgriLifeExtension.tamu.edu; Masoner JR, Mladinich CS, Konduris AM, Smith SJ. 2003. Comparison of irrigation water use estimates calculated from remotely sensed irrigated acres and state reported irrigated acres in the Lake Altus drainage basin, Oklahoma and Texas, 2000 growing season. Water-Resources Investigations Report 03-4155, U.S. Geological Survey: Oklahoma City, OK.

Statistics on water use required to produce different products are in: US Environmental Protection Agency. 2012. Water trivia facts. http://water.epa.gov/learn/kids/drinkingwater/water_trivia_facts.cfm#_ednref10 (downloaded Dec 27, 2012).

The problems created by fish invasions in rivers, and illegal introductions are reviewed in: Fausch KD, Garcia-Berthou E. 2013. The problem of invasive species in river ecosystems. In Sabater S, Elosegi A, editors. *River Conservation: Challenges and Opportunities*, 193–216. BBVA Foundation. Bilbao, Spain; Johnson BM, Arlinghaus R, Martinez PJ. 2009. Are we doing all we can to stem the tide of illegal fish stocking? *Fisheries* 34:389–394.

The potential effects of introduced hatchery fish, and other human actions, on Columbia River food webs are presented in: Naiman RJ, Alldredge JR, Beauchamp DA, Bisson PA, Congleton J, Henny CJ, Huntly N, Lamberson R, Levings C, Merrill EN, Pearcy WG, Rieman BE, Ruggerone GT, Scarnecchia D, Smouse PE, Wood CC. 2012. Developing a broader scientific foundation for river restoration: Columbia River food webs. *Proceedings of the National Academy of Sciences USA* 109:21201–21207.

Prevalence and effects of altered flow regimes are described in: Poff NL, Olden JD, Merritt DM, Pepin DM. 2007. Homogenization of regional river dynamics by dams and global biodiversity implications. *Proceedings of the National Academy of Sciences USA*. 104:5732–5737; Poff NL, Richter BD, Arthington AH, Bunn SE, Naiman RJ, Kendy E, Acreman M, Apse C, Bledsoe BP, Freeman MC, Henriksen J, Jacobson RB, Kennen JG, Merritt DM, O'Keefe JH, Olden JD, Rogers K, Tharme RE, Warner A. 2010. The ecological limits of hydrologic alteration (ELOHA): A new framework for developing regional environmental flow standards. *Freshwater Biology* 55:147–170.

Effects of urbanization on streams is described in: Walsh CJ, Roy AH, Feminella JW, Cottingham PD, Groffman PM, Morgan RP. 2005. The urban stream syndrome: Current knowledge and the search for a cure. *Journal of the North American Benthological Society* 24:706–723.

Gene Odum's theory on an appropriate mix of environments to sustain societies is from: Odum EP. 1969. The strategy of ecosystem development. *Science* 164: 262–270.

The idea of rivers as integrated ecological and socioeconomic systems is from: Naiman, R.J. 2013. Socio-ecological complexity and the restoration of river ecosystems. *Inland Waters*, 3: 391–410; Rieman BE, Smith CL, Naiman RJ, Ruggerone GT, Wood CC, Huntly N, Merrill EN, Alldredge JR, Bisson PA, Congleton J, Fausch KD, Lamberson R, Levings C, Pearcy W, Scharnecchia D, Smouse P. In press. A comprehensive approach for habitat restoration in the Columbia basin. *Fisheries*.

Methods to prevent sediment from draining into streams are described in: Waters TF. 1995. Sediment in streams: Sources, biological effects and control. *American Fisheries Society Monograph* 7. Bethesda, MD (see p. 151).

The number of headwater tributaries to the Cache la Poudre River (about fifth order) was calculated from Horton's Law of Stream Numbers, based on a bifurcation ratio of 3.5. See: Leopold LB, Wolman MG, Miller JP. 1964. *Fluvial Processes in Geomorphology*. W. H. Freeman and Co.: San Francisco, CA.

The idea of not letting what we cannot do prevent us from doing what we can do is after a quote by Coach John Wooden. See ESPN.com Staff writers. 2010. *The Wizard's Wisdom: "Woodenisms"* (Quotes attributed to John Wooden). http://sports.espn.go.com/ncb/news/story?id=5249709 (downloaded December 27, 2012).

Research on human preference for the sound of running water in urban areas and the research on sound in Swiss rivers are summarized in: Carles JL, Barrio IL, Vincent de Lucio J. 1999. Sound influence on landscape values. *Landscape and Urban Planning* 43:191–200; Tonella D, Tockner K. 2012. Research on the sound aesthetics in Swiss rivers. Unpublished abstract. Personal communication. July 10, 2012.

Research by Emily Bernhardt and Margaret Palmer on river restoration is presented in: Bernhardt ES, Palmer MA, Allan JD, Alexander G, Brooks S, Carr J, Dahm C, Follstad-Shah J, Galat DL, Gloss S, Goodwin P, Hart D, Hassett B, Jenkinson R, Kondolf GM, Lake S, Lave R, Meyer JL, O'Donnell TK, Pagano L, Srivastava P, Sudduth E. 2005. Restoration of U.S. Rivers: A national synthesis. *Science* 308:636–637; Bernhardt ES, Palmer MA. 2011. River restoration: The fuzzy logic of repairing reaches to reverse catchment scale degradation. *Ecological Applications* 21:1926–1931; Palmer MA. 2009. Reforming watershed restoration: Science in need of application and applications in need of science. *Estuaries and Coasts* 32:1–17; Palmer MA, Menninger HL, Bernhardt ES. 2010. River restoration, habitat heterogeneity and biodiversity: A failure of theory or practice? *Freshwater Biology* 55:205–222.

An example of successful restoration, and a scientific theory of restoring processes are presented in: Beechie TJ, Sear DA, Olden JD, Pess GR, Buffington JM, Moir H, Roni P, Pollock MM. 2010. Process-based principles for restoring river ecosystems. *BioScience* 60:209–222; White SL, Gowan C, Fausch KD, Harris JG, Saunders WC. 2011. Response of trout populations in five Colorado streams two decades after habitat manipulation. *Canadian Journal of Fisheries and Aquatic Sciences* 68:2057–2063; Roni P, Beechie T, editors. 2013. *Stream and Watershed Restoration: A Guide to Restoring Riverine Processes and Habitats*. Wiley-Blackwell: Chichester, UK.

Key ideas on the importance of natural disturbances in sustaining biological communities and ecosystems are in: Connell JH. 1978. Diversity in tropical rainforests and coral reefs. *Science* 199:1302–1310; Huston M. 1979. A general hypothesis of species diversity. *American Naturalist* 113:81–102; Pickett STA, White PS, editors. 1987. *The Ecology of Natural Disturbance and Patch Dynamics*. Academic Press: New York, NY.

Research by Mary Power on the importance of flooding disturbance to river food webs is in: Power ME. 1992. Hydrologic and trophic controls of seasonal algal blooms in northern California rivers. *Archiv für Hydrobiologie* 125:385–410; Wootton JT, Parker MS, Power ME. 1996. Effects of disturbance on river food webs. *Science* 273:1558–1560; Power ME, Parker MS, Dietrich WE. 2008. Seasonal reassembly of a river food web: Floods, droughts, and impacts of fish. *Ecological Monographs* 78:263–282.

Listing of California steelhead trout under the Endangered Species Act is available at the NOAA Fisheries website: http://www.nwr.noaa.gov/ESA-Salmon-Listings/Salmon-Populations/Steelhead/ (accessed Jan 21, 2013).

The role of natural disturbances from landslides and fires in shaping stream habitat is described in: Reeves GH, Benda LE, Burnett KM, Bisson PA, Sedell JR. 1995. A disturbance-based ecosystem approach to maintaining and restoring freshwater habitats of evolutionarily significant units of anadromous salmonids in the Pacific Northwest. *American Fisheries Society Symposium* 17:334–349; Rieman B, Clayton J. 1997. Wildfire and native fish: issues of forest health and conservation of sensitive species. *Fisheries (Bethesda)* 22(11): 6–15; Dewberry C, Burns P, Hood L. 1998. After the flood: The effects of the storms of 1996 on a creek restoration project in Oregon. *Restoration and Management Notes* 16:174–181.

The proposal for diverting and storing peak flows from my local river is in: U.S. Army Corps of Engineers. 2008. Draft Environmental Impact Statement: Northern Integrated Supply Project for the Northern Colorado Water Conservancy District. USCOE Omaha District: Omaha, NE.

Results of large-scale experiments on the effects of restored flow regimes in rivers are reviewed in: Konrad CP, Olden JD, Gido KB, Hemphill NP, Kennard MJ, Lytle DA, Melis TS, Robinson CT, Schmidt JC, Bray EN, Freeman MC, McMullen LE, Mims MC, Pyron M, Williams JG. 2011. Large-scale flow experiments for managing rivers. *BioScience* 61: 948–959; Konrad CP, Warner A, Higgins JV. 2011. Evaluating dam re-operation for freshwater conservation in the Sustainable Rivers Project. *River Research and Applications* 28:777–792; Robinson CT. 2012. Long-term changes in community assembly, resistance, and resilience following experimental floods. *Ecological Applications* 22:1949–1961; Cross WF, Baxter CV, Rosi-Marshall EJ, Hall RO Jr, Kennedy TA, Donner KC, Seegert SEZ, Kelly HAW, Yard MD, Behn KE. 2013. Foodweb dynamics in a large river discontinuum. *Ecological Monographs* 83:311–337.

Restoring habitat in Pacific Northwest rivers by mimicking the processes and locations of landslides, and the theory behind process restoration, are described in: Reeves GH, Benda LE, Burnett KM, Bisson PA, Sedell JR. 1995. A disturbance-based ecosystem approach to maintaining and restoring freshwater habitats of evolutionarily significant units of anadromous salmonids in the Pacific Northwest. *American Fisheries Society Symposium* 17:334–349; Dewberry C, Burns P, Hood L. 1998. After the flood: The effects of the storms of 1996 on a creek restoration project in Oregon. *Restoration and Management Notes* 16:174–181; Beechie TJ, Sear DA, Olden JD, Pess GR, Buffington JM, Moir H, Roni P, Pollock MM. 2010. Process-based principles for restoring river ecosystems. *BioScience* 60:209–222.

Ideas from Heraclitus are quoted by Plato in the dialogue Cratylus (http://en.wikiquote.org/wiki/Heraclitus; accessed Jan 21, 2013), and recorded in Harris (no date).

Kahlil Gibran's poem is in: Gibran, K. 1923. On children. In *The Prophet*, 17–18. Alfred A. Knopf: New York, NY.

Ideas on conserving what we love, and Leopold's "round river" are found in: Leopold A. 1953. Conservation. In Leopold LB, editor. *Round River: From the journals of Aldo Leopold*, 145–157. Oxford University Press: Oxford, UK; Leopold A. 1953. Round River. In Leopold LB, editor. *Round River: From the journals of Aldo Leopold*, 158–165. Oxford University Press: Oxford, UK; Gould SJ. 1991. Unenchanted evening. *Natural History* 100(9): 4–10; Orr DW. 2010. Love. In *Hope is an imperative: The essential David Orr*, 30–34. Island Press: Covelo, CA.

The famous quote by Baba Dioum reads: "In the end, we will conserve only what we love. We will love only what we understand. We will understand only what we are taught." See: Rodes BK, Odell R. 1997. *A Dictionary of Environmental Quotations*. The Johns Hopkins University Press: Baltimore, MD.

One of the first instances where Aldo Leopold used the ecosystem concept is in: Leopold A. 1939. A Biotic View of Land. In Flader SL, Callicott JB, editors. *The River of the Mother of God, and Other Essays by Aldo Leopold*, 266–273. The University of Wisconsin Press: Madison.

The quote from Alexis de Tocqueville is from: de Tocqueville, Alexis. 2000. *Democracy in America*, 232. Edited and translated by Mansfield HC, Winthrop D. University of Chicago Press: Chicago, IL.

Bill McKibben's observations are in: McKibben, B. 2010. *Eaarth: Making a Life on a Tough New Planet*, 206–212. Times Books. Henry Holt and Company: New York, NY.

Aldo Leopold's quote about goose music is from: Leopold A. 1953. Goose Music. In Leopold LB, editor. *Round River: From the journals of Aldo Leopold*, 166–173. Oxford University Press: Oxford, UK.

Notes to Epilogue: Nakano's Legacy

Nakano's paper from his early research in this pool is in: Nakano S. 1995. Competitive interactions for foraging microhabitats in a size-structured interspecific dominance hierarchy of two sympatric stream salmonids in a natural habitat. *Canadian Journal of Zoology* 73:1845–1854.

Recent reviews and research on linkages between streams and riparian zones, spurred by the seminal work of Nakano and others includes: Baxter CV, Fausch KD, Saunders WC. 2005. Tangled webs: Reciprocal flows of invertebrate prey link streams and riparian zones. *Freshwater Biology* 50:201–220; Saunders WC, Fausch KD. 2007. Improved grazing management increases terrestrial invertebrate inputs that feed trout in Wyoming rangeland streams. *Transactions American Fisheries Society* 136:1216–1230; Walters DM, Fritz KM, Otter RR. 2008. The dark side of subsidies: Adult stream insects export organic contaminants to riparian predators. *Ecological Applications* 18:1835–1841; Sato T, Watanabe K, Kanaiwa M, Niizuma Y, Harada Y, Lafferty KD. 2011. Nematomorph parasites drive energy flow through a riparian ecosystem. *Ecology* 92:201–207; Sato T, Egusa T, Fukushima K, Oda T, Ohte N, Tokuchi N, Watanabe K, Kanaiwa M, Murakami I, Lafferty KD. 2012. Nematomorph parasites indirectly alter the food web and ecosystem function of streams through behavioural manipulation of their cricket hosts. *Ecology Letters* 15:786–793; Saunders WC, Fausch KD. 2012. Grazing management influences the subsidy of terrestrial prey to trout in central Rocky Mountain streams (USA). *Freshwater Biology* 57:1512–1529; Kraus JM, Schmidt TS, Walters DM, Wanty RB, Zuellig RE, Wolf RE. 2014. Cross-ecosystem impacts of stream pollution reduce resource and contaminant flux to riparian food webs. *Ecological Applications* 24:235–243.

I addressed Hiroya Kawanabe here as *sensei* (teacher), with the deepest respect.

The history and meaning of the shrines associated with water around Lake Biwa, and the Kibune Shrine in particular, is from: Kawanabe H. 2003. Cultural associations in an ancient lake: Gods of water in Lake Biwa and the River Yodo basin, Japan. *Hydrobiologia* 500:213–216; Anonymous. The water god: Kifune Shrine (Kifune Jinja). Pamphlet provided at Kifune Shrine, August 1, 2012. See also www.kibune.jp/jinja/ [Note: Kifune and Kibune are interchangeable, because early Japanese language did not distinguish between these English consonants. H. Kawanabe, personal communication.]

Tokichi Kani's stream classification is described in: Kani T. 1944. Ecology of torrent-inhabiting insects. In Furukawa H, editor. *Insects 1*, 171–317. Kenkyu-sha Press: Tokyo, Japan (in Japanese). Cited in: Mizuno N, Kawanabe H. 1981. A topographical classification of streams, with an introduction of the system widely used in Japan. I. Reach type, stream zone and stream type. *Verhandlungen Internationale Vereinigung für Theoretische und Angewandte Limnologie* 21:913.

Kani's influence on other early Japanese ecologists is described in: Fausch KD, Nakano S. 1998. Research on fish ecology in Japan: A brief history and selected review. *Environmental Biology of Fishes* 52:75–95.

Index

S

T

Kurt Fausch is a professor in the Department of Fish, Wildlife, and Conservation Biology at Colorado State University, where he has taught for 33 years. His collaborative research has taken him throughout Colorado and the West, and worldwide, including to Hokkaido, Japan, where his experiences were chronicled in the PBS documentary *RiverWebs*. He has received awards from the American Fisheries Society and the World Council of Fisheries Societies, and served as the acting director of the Graduate Degree Program in Ecology at CSU.